Advanced

Bee - Culture

Its

Methods and Management

BY

W. Z. HUTCHINSON

Editor Bee-Keepers' Review

Flint, Michigan

Third Edition

1905

Advanced Bee-Culture

ISBN 978-1-912271-06-1

Published by Northern Bee Books, 2017
Scout Bottom Farm
Mytholmroyd
Hebden Bridge HX7 5JS (UK)

Printed by Lightning Source, UK

DEDICATION.

To those who are getting their bread and butter by raising honey to spread upon the bread and butter of others, this book is dedicated by

The Author.

Introduction.

THIS book is intended for the professional bee-keeper; hence it is taken for granted that the reader is fairly well acquainted with bees and their management.

No space is devoted to the natural history, anatomy and physiology of the bee, because my experience has been along commercial, instead of scientific, lines.

The history of this book, how and why it came to be written and published, would read something as follows: Twenty or more years ago, while making my living in the apiary, I learned that, with my management, it was more profitable to use starters, only, in the brood-nest when hiving swarms. My experiments and methods were described in print; others tried my plans; discussions followed; and, finally, it became apparent that the system was really more complex than it appeared on the surface; also, that short articles scattered through different bee journals did not present the subject in the best possible manner, and, as a result, I published a little book in which I described, in detail, my method of comb honey production.

One of the criticisms brought against the little book was its small size; and I was repeatedly urged to write a larger book, giving my experience and views more in detail, and upon other points. Flattering as all of this may have been, I doubt if I should have yielded to these entreaties had it not been that by the time the last copy of the little book was sold, I had been editor of the Bee-Keepers' Review for nearly four years, and had the benefit of reading, and studying over, special discussions, by the most practical men, of the most important questions connected with our pursuit. As it was, I went to work and classified, arranged and condensed, and gave what I considered the cream of the special topic discussions that had appeared in the Review. So many new subjects were taken up that the old title, "The Production of Comb Honey," was no longer appropriate, and, as I was giving what seemed to me the best and most advanced methods, I called the new book ADVANCED BEE CULTURE. Two years ago, the first edition having been exhausted, I re-wrote and revised everything necessary to bring it up to date, and got out a second edition which has since been sold. The present edition has been largely re-written; many engravings, much new matter, and a more substantial binding, being added; thus bringing the book more nearly up to the ideal that I have for several years had in mind.

ADVANCED BEE CULTURE is really the summing up of the best that has appeared in the Bee-Keepers' Review during the 18 years of its existence; that is, from a most careful examination of the views of the most progressive men, and a thorough consideration of the same in the light of my experience as a bee-keeper, I have described in plain and simple language what I believe to be the most advanced methods of managing bees, *for profit*, from the beginning of the season throughout the entire year.

W. Z. HUTCHINSON, Flint, Mich.

As ever yours,
W. Z. Hutchinson.

Bee-Keeping as a Business.

IN reply to the query, "What will best mix with bee-keeping?" I have always replied: "Some more bees." When the conditions are favorable, I am decidedly in favor of bee-keeping as a specialty—of dropping all other hampering pursuits, and turning the whole capital, time and energies into bee-keeping. If bee-keeping cannot be made profitable as a specialty, then it is unprofitable as a subsidary pursuit. If bee-keeping must be propped up with some other pursuit, then we better throw away bee-keeping, and *keep the prop.*

General farming is very poorly adapted for combining with bee-keeping, yet the attempt is probably made oftener than with any other pursuit. There are critical times in bee-keeping that will brook no delay, when three or four days or a week's neglect may mean the loss of a crop; and these times come right in the height of the season, when the farmer is the busiest. Leaving the team and reaper standing idle in the back field while the farmer goes to the house to hive bees, is neither pleasant nor profitable. Drawing in a field of hay, while the bees lie idle because the honey has not been extracted to give them store-room, is another illustration of the conditions with which the farmer-bee-keeper has to contend. The serious part of it is that the honey thus lost may be worth nearly or quite as much as the hay that is saved. Some special lines of rural pursuits, like winter-dairying or the raising of grapes, or winter-apples, unite with bee-keeping to much better advantage than general farming; but when bee-keeping is capable of absorbing all of the capital, time and energy that a man can put into it, why divide these resources with some other pursuit? It has been said that bee-keeping is a precarious pursuit, that it cannot be depended upon, alone, to furnish a livelihood; and, for this reason, it should be joined with some business of a more stable character. It is true that there are many localities where there is often a season in which little or no honey is secured, and, in the Northern States, winter-losses are

Hard Maple Forests of Northern Michigan.

As fast as the timber is lumbered off, red raspberries spring up in myriads, furnishing bee
pasture that is simply incomparable.

sometimes very heavy, hence it would be risky to depend entirely for a living upon keeping bees, in a *limited* way, in such localities; but, if the average profit from bee-keeping, one year with another, is not the equal of other rural pursuits, why keep bees? The truth of the matter is that it is greater; and if bee-keepers would only drop everything else, and adopt methods that would enable them to branch out and keep hundreds of colonies where they now have dozens, they would secure enough honey in the good years to more than carry them over the poor years, and thus not only make a living, but lay up money.

When a man decides to cut loose from everything else, and go into bee-keeping extensively, making it his only and his life-business, the question of all questions is that of locality. There are few localities in which a small apiary might not yield some surplus, but when a man is to make of bee-keeping his sole business, the securing of the best possible location is time and money well spent. What a good, solid foundation is to a "sky-scraper," a good location is to the building up of a successful, extensive bee business. Having settled in a locality, the bee-keeper can not study it too thoroughly. Especially must he understand its honey resources; the time when each flow begins, its probable duration, its quantity and character. He must know whether to expect a spring-flow, like that from dandelion, hard maple, or fruit-bloom, that will build up the colonies for the main harvest that is to come later. If there is likely to be a season of scarcity between the early flow and the main harvest, it must be known, and preparations made to keep up brood rearing by means of feeding or the uncapping of honey. The management will depend largely upon the source of the main honey-flow, whether it be raspberry, clover, basswood, buckwheat, alfalfa, sage, or fall flowers. Whatever the source, the bee-keeper must know when to expect it, and plan to have his colonies in exactly the right condition to gather it when it comes. This is one of the fundamental principles of successful bee-keeping.

Having secured the most desirable location, the next step is to procure the best kind of bees that can be obtained. There are several different varieties of bees, each with its peculiarities, but, aside from this, every bee-keeper who has had experience with several strains of the same variety, knows that some strains are far superior to others—that there is scrub-stock among bees, just as there are scrub-horses, cattle, sheep and poultry. With scrub-stock, the cost of hives, combs and other appliances remains the same; it is no less work to care for such stock; and it requires the same amount of honey to raise and feed it as it does the best stock in the world. In

Apple Orchard in Full Bloom.

There is nothing in the line of early honey that so stimulates brood rearing as does that from the pink and white blossoms of the apple trees.

proportion to its cost, no investment brings the bee-keeper greater profit than the securing of superior stock.

Having secured a good location, and good stock, the bee-keeper should adopt such hives, implements and methods as will enable him to branch out, establish out-apiaries, and keep a large number of colonies. At the present time the greatest failing of professional bee-keepers is the keeping of too few bees—of clinging to some other hampering pursuit. Many keep enough bees to furnish them a fair living in a good season, but when winter losses, and poor honey seasons follow one another in quick succession, there is suffering; or, at least, great inconvenience. If a man is going to follow bee-keeping as a profession, his only hope is in a good location, good stock, and the keeping of bees in such numbers that when a good year comes he can pile up the honey ton upon ton—enough to keep him several years. The larger a business the more cheaply can it be conducted in proportion to the results; not only this, but the very fact that bees are scattered about in out-apiaries several miles apart, adds to the certainty of the crop; as one locality often yields a fair crop while another a few miles away yields nothing.

It has been urged against bee-keeping as a sole pursuit that, while it keeps a man very busy during the summer it leaves him idle in the winter. Bee-keeping, rightly managed, will keep a man busy every day in the year. Too many bee-keepers fail to realize that the selling of a crop is fully as important as its production. The business part of bee-keeping has been sadly neglected. No set rule can be given as to how a man shall dispose of his crop, but it does seem like very poor business management to send away a crop of honey to some commission merchant, and then sit around all winter when good wages might be made selling honey direct to consumers, or to retail dealers. The selling of the crop, and the preparations for the coming season, may well occupy a man during the winter.

It should be understood, however, that bee-keeping is not an occupation in which one can easily become *wealthy*. In this respect, it is much like other rural pursuits. Rightly managed, in a locality adapted to the business, it can be depended upon to furnish a comfortable living, and perhaps enable a man to lay up a few thousands of dollars, but such fortunes as are sometimes amassed in merchandising or manufacturing can never be hoped for by the bee-keeper. Fortunately, however, the perfection of a man's happiness bears but little relation to the size of his fortune; and many a man with the hum of the bees over his head, finds happiness deeper and sweeter than ever comes to the merchant prince with his cares and his thousands.

One of Six Out-Apiaries Belonging to M. A. Gill of Colorado.

Making a Start in Bee-Keeping.

BEES, and the requisite knowledge for their management, are the two most important factors in making a start in the business. The latter ought to be secured first; or, at least, its acquisition should keep pace with any increase in numbers of the former. As in any business, so with bee-keeping, it pays well to lay broad and deep the foundations of an education in that line of work. So many men fail, in different kinds of business, because they start in with only a narrow or superficial knowledge of their chosen profession. The time may come when bee culture will be taught at the agricultural colleges, the same as dairying is now taught, but, at present, the nearest approach to a college-course, is that of working with some experienced, successful bee-keeper. This is the quickest way of learning bee-keeping; and, if the teacher is competent, is a very desirable method. The beginner is not always able to choose wisely in selecting an instructor, hence, it is well to supplement such instruction by a course of reading, and thus be able to make comparisons, and discuss the instructor's methods in the light of those employed by others. In fact, I am inclined to think that a thorough course of reading is the most desirable *first* step that can be taken by a prospective bee-keeper. One after the other, I would read the leading, standard text books. Having done this, the next step is to subscribe for the best bee journals. At this stage, a season with an expert bee-keeper would be of great value; when the previous reading will enable the apprentice to intelligently use his mind, and see the *reason* of things, instead of being simply an imitator, following blindly in the footsteps of his preceptor. Probably nine-tenths of the men who now keep bees, never served an apprenticeship. Many have become interested in bees from the capture of a stray swarm. Neighboring bee-keepers would be visited, books or papers borrowed or bought, improved hives and methods adopted, and, as the bees increased, so did the enthusiasm and interest, until, finally, the bees received more time and attention than did the regular business; thus did bee-keeping eventually become a specialty or the sole business.

When a man has decided to embark in bee-keeping as a business, he should in some manner learn the business thoroughly before investing extensively. No hard and fast rules can be laid down; so much depending upon circumstances. A young man with no established business, would do well to pass one or two seasons in the employ of some experienced bee-keeper, as has been already suggested, while an older man already in business, with a family to support, may find it advisable to gradually work into bee-keeping, reading and studying as his bees increase. Whatever the method employed, let the work be thorough; and, especially, let there be plenty of *actual experience* before venturing extensively.

As a rule, a man already has some bees when he decides to become a bee-keeper. Perhaps he never formally makes any such decision. He captures a stray swarm, or cuts a bee tree and saves the bees, and the stock increases with such wonderful rapidity that the owner becomes a bee-keeper ere he scarcely realizes it. This wonderful rapidity with which bees increase is one strong argument in favor of a man securing a few colonies and building them up into an apiary, instead of buying a large number of colonies at the beginning. By rearing queens with which to furnish the newly made colonies, and furnishing them with full sheets of comb foundation, the extent to which bees can be increased in a favorable season is something almost beyond belief. Just how, or where, the first colonies shall be secured may well be considered. As a rule, the man who has steady work, at good wages, better buy bees in such movable comb hives as he intends to use. If he can get them near home, of some reliable bee-keeper, so much the better. Of course, there are instances in which a man has more time than money, or there may be a trace of the sportsman in his make-up, and, in either case, the hunting of bees, or the putting out of decoy hives to catch stray swarms, will make to him a strong appeal. In those parts of the country where many bees are kept, yet there is not much timber, as in Colorado or California, there is no difficulty in catching swarms in decoy hives; in fact, there is difficulty in keeping swarms out of chimneys, and the walls of buildings. While out riding one day with Mr. Gill, of Colorado, he pointed out one house in the walls of which five colonies had their homes. In California Mr. Mendelson set away three empty hives in his wagon shed, and when I was there swarms had taken possession of two of the hives. Even in well-timbered localities, stray swarms are frequently caught in decoy hives. Mr. George A. Fenton, of Pine Island, Minnesota, reported in the Review, in 1900, that he had, the previous year, caught more than 50 swarms in decoy hives. He used ordinary boxes instead of hives, and put them pretty high up in tall trees, as a good hive, easily accessible, is quite likely to be stolen. A piece of old, black comb is fastened inside the hive, or box, and the latter firmly fastened to the tree so as not to be easily blown down, a position being chosen where the hive will be in the shade. A tree on the edge of the woods is chosen, as, when a swarm reaches the woods it at once begins a search for a suitable hollow in which to make its home. The hives are examined as often as once a week, oftener if there is time, and when one is found occupied the tree is ascended by the use of climbers, the box lowered by means of a rope, and another box fastened up in its place. Mr. Fenton attributed his success to the fact that the farmers near him who had a few bees did not watch them, but let the swarms go to the woods, and, further, there was no *large* timber for many miles around.

In all probability, more of my readers would be able to secure bees by finding and cutting bee-trees, than by putting out decoy hives, and, as the subject will not be touched upon elsewhere, I will

DECOY HIVE IN POSITION.

here describe how bee-trees are found. In principle, it is simply that of putting out honey, in times of scarcity, when bees will "rob," watching the loaded bees as they fly home, and following the "line" of bees to the tree, but there are many details the observance of which greatly aids the hunter in his search. Those who hunt bees to any great extent use what is called a "bee-box." This is a small

"Lining" Wild Bees.

box made in two parts, the lower half being used to hold a piece of comb filled with honey, while the upper part, or cover, is used principally for catching the bees and getting them at work upon the "bait." The top of the upper part is covered with a glass, and a short distance below the glass is a horizontal, sliding partition; while still lower, just at the lower edge of one of the sides, is a small opening covered on the inner side with glass. Equipped with his box and a bottle of diluted honey, the bee hunter begins operations in some field or cleared spot near the forest in which he expects bee-trees may be found. The honey is diluted with an equal amount of water, as it enables the bees to load and unload quicker, to fly faster, and in a more direct line. Sometimes pieces of old comb are burned, the odor from the "smudge" attracting bees from a long distance. If, by careful search, a bee is found industriously at work upon some weed, the cover to the box is taken off, the slide drawn nearly out, and the open or lower side of the cover held near the bee. A hand-kerchief is then held upon the opposite side of the bee, and, as the cover and the handkerchief are brought quickly together, the bee is caught in the former. Seeing the light, the bee at once buzzes up against the glass top of the cover, when the slide is shoved in, thus making the bee a prisoner. The cover is now replaced upon the box, the box set upon a stump or upon a stake stuck in the ground, the slide drawn nearly out, and the handkerchief spread over the glass top. The bee now sees only one opening, the small one in the side of the cover near its lower edge, and in attempting to escape by the lower opening, the bee comes in contact with the comb of honey in the lower part of the box.

To find the honey is to at once begin "loading up." Occasionally removing the handkerchief shows when the bee has found the honey, and as soon as it is seen filling its sac, the hunter carefully removes the cover, and places his eye near the ground. This position is assumed to secure the sky as a background in watching the bee take its homeward flight. Under such conditions a bee can be kept in sight for a long distance. A minute or two suffices for the bee to fill·its honey sac, when it slowly rises in gradually widening circles. Each time around it sways more and more to one side—toward the spot where it lives; finally, having taken its "bearings," it strikes a "bee-line" for home. In a short time it returns with perhaps three or four companions in its wake—eager to learn from whence came that fine load of honey. The result is that a strong "line" of bees is finally at work between their home and the hunter's box. He now puts the cover on the box, shutting in the bees, and moves along on the "line" towards their home. After going some distance the bees are released, when they at once leave for home, only to return and re-establish the "line," when the hunter again closes the box and moves forward. When the bees turn and fly back on the line, it shows that the tree has been passed and must be near at hand. At this point in the game it may be advisable to resort to what is known as "cross-lining," that is, the box is moved off several rods to one side, and another, or "cross-line," established. The tree must certainly be near the point where these two lines intersect. The trunks and branches of all large trees in the vicinity are now carefully examined, particular attention being paid to any knot holes or openings. Getting the tree between the sun and the observer greatly aids in discovering any bees that may be flitting about. An opera glass is also a great aid in this part of the work.

Bee-trees are also found by walking through the woods in the first warm days in the spring, before the snow is off the ground, listening to their humming and noticing the dead bees that have been brought out and dropped upon the snow.

After the bees have been found, then comes the task of getting them out of the tree and into a hive. Sometimes it is possible, if they are located in a large limb, to cut off the limb beyond the portion occupied by the bees, and then cut off the portion in which they are located, and lower it by means of a rope. Again, it is sometimes possible to rig up a temporary scaffold, and cut out a piece of the tree over the bees' home, remove the combs and lower them in a basket. Such proceedings are attended with more or less danger, even when carried out by the most careful of men, and I would rather put up with the more or less broken condition of the combs that usually result from cutting down the tree, than with broken legs or arms. Many times a tree can be so guided that it will strike upon smaller trees that will break the fall. The saving of the bees and combs, after they have been reached, is very similar to an ordinary job of transferring from an old box hive to a movable comb hive. Pieces of comb containing brood must be fastened into frames, and hung in the hive, and as many as possible of the bees guided into the entrance. If the hive is left on the spot for several hours, perhaps over night, nearly all of the live bees will gather into the hive.

As I said at the beginning, if a man has steady work at good wages, he will, as a rule, find it more satisfactory to buy bees in good, movable comb hives; but, if he has the time and inclination to get a start by hunting bees, or by putting up decoy hives, what I have written will show him how to do it.

Mistakes in Bee-Keeping.

IT is pleasant to tell of success. Mistakes are mentioned with reluctance; yet they may be of equal value for imparting information. Mr. J. M. Smith of Wisconsin is a noted horticulturist. The crops of berries and cabbages that he raises are something wonderful. His contributions to the press are valuable; but I never read one that contained more information than the one in which he recounted the *mistakes* of his horticultural life. I believe that space can be profitably used in mentioning a few things that experienced bee-keepers look upon as mistakes in bee-keeping.

A man who has *decided* that he will make bee-keeping his life-business, makes a mistake when he gets a few colonies and attempts to learn the business all by himself. Both time and money will be saved by passing at least one season in the employ of some successful bee-keeper.

If a man must start with a few colonies, and learn the business by himself, let him avoid the mistake of attempting to follow several leaders or systems. Much confusion and annoyance will be saved if he adopts the teachings, methods and appliances of some one successful bee-keeper. He may make the mistake of not choosing the best system, but better this than a mixture of several systems.

A beginner is quite likely to fall into the error of increasing his colonies too rapidly. There is probably no mistake so disastrous as this, on account of its frequency and results. To the beginner, this is very tempting ground. If bee-keeping must be learned by experience and reading (without the serving of an apprenticeship) the beginning should be small, and practical knowledge and skill keep pace with the increase of colonies.

A mistake that has been made by many is in looking upon bee-keeping as a sort of royal road to wealth, or, at least, a good living, with but little labor, and, some believe, little brains, after they have once "caught on" to a few secrets. (?) To choose any business simply because it is profitable, is the height of folly. A business that is unusually profitable does not long remain such. It soon becomes overcrowded, and loses its bonanza character. A man should choose a business because he and his surroundings are best adapted to the pursuit.

Many fall into the error of judging entirely by *results*, regardless of causes. As that excellent bee-keeper, R. L. Taylor, of Michigan, once said: "The greatest actual results do not prove the method of management by which they were produced to be the best. Time, and labor, and thought, and care, and material, and capital, are all money, so the greatest results numerically, may be obtained at a loss, while the least apparent results may yield a profit."

In much this same manner do many bee-keepers make the mistake of computing their income at so many pounds per colony, and at so much per pound. The greatest yield per colony might nct be so profitable as a less yield per colony from more colonies, or even a lessened yield from the same number of colonies. If a great yield per colony is the result of a great deal of work, it may be that the work was done at a loss. Bee-keeping should be viewed in a broader light. It may sometimes be profitable to put a great deal of work on each colony, but each bee-keeper should ask himself, how, *all* things considered, can I make the most profit ? That is the question, and all other propositions not relating directly thereto are mistakes.

And this leads to the mention of another mistake, the keeping of too few bees. Instead of keeping only a few colonies, and striving to secure the largest yield per colony, it is usually more profitable to keep more bees—enough to gather all the honey in a given area, and then when that area is overstocked, it is probably a mistake not to start out-apiaries. There is much to be gained in having as few *kinds* of things to do as possible, and as *much* of them as can be managed. The proportional cost of doing business is greatly lessened by increasing the volume.

Another mistake is that of choosing hives, implements and methods that are complicated and require much time for their manipulation. A most common error in this direction is that of trying to adapt hives to *bees*, to such an extent as to almost entirely ignore the adaptability of the hive to the bee-keeper. I remember once hearing a bee-keeper arguing for a hive that it was "so handy for the bees." "Why," said he, "if you were building a house, would you have it so arranged that your wife would be compelled to go up and down stairs between the kitchen and the pantry?" It must be remembered that we build hives for our bees, and houses for our wives, with altogether different objects in view. We don't keep bees, nor arrange their hives, so much with a view to saving them labor, as that *we* may get the most honey with the least labor to *ourselves*. Drone-traps, queen-traps, self-hivers, queen-excluders, separators, and many other contrivances, are probably not considered "handy" by the bees, but their use is an advantage to us.

It is in line with this method of reasoning that causes some bee-keepers to make the mistake of condemning any practice that is not "according to nature." The whole system of modern bee culture is largely a transgression of nature's laws, or of managing differently than the bees would manage if left to their own way of doing things. In some things it is advisable to allow nature to have her own way, in others it is not, and our success is just in proportion as we learn when and where we can, advantageously, to a certain extent, cross nature's methods with those of man's intelligence.

Mistakes have been made, and erroneous conclusions arrived at, by experimenting upon too small a scale. There are some kinds of experiments which will demonstrate truths just as well upon a small as upon a large scale, while there are others requiring experiments upon a large scale, and a repetition of experiments, before definite conclusions can be arrived at.

Many beginners make the mistake of thinking that they can improve some of the standard hives and implements; and that before they have fairly learned the business. A beginner with a few brains, boards, and buzz saw, is the man of all others who feels called upon to invent a bee hive.

Others make the mistake of adopting new hives, implements, methods, or varieties of bees, upon too large a scale before they are certain that the change will be desirable. When a new thing with one advantage is held up before our eyes, we are too much given to forgetting the many advantages possessed by the article that we are asked to lay aside for the new comer. As a rule, the rank and file can afford to wait until at least good reports are given in regard to a novelty. Then it will be in order to experiment upon no larger scale than that upon which failure can be met and borne.

Speaking of the "rank and file" waiting for the leaders or others to try novelties, reminds me that it is a mistake to have *undue* confidence in the leaders, or in any one, for that matter. It is possible that they may be in error, or some unknown circumstances may cause different results at different times in other localities. It is a mistake to pin one's faith blindly to another. Read how other men have succeeded, consider their advice, but do your own thinking just the same, and try things for yourself until you are sure you are right, then go ahead.

One expensive mistake, yet one that is easily avoided, is made year after year by many bee-keepers, and that is not securing hives, sections, foundation and other supplies in season. They *intend* to buy them soon enough, but wait until the last moment. So many others do the same thing that dealers and manufacturers are over-run with orders, and expensive and vexatious delays occur. A delay of a few days, at just the right time, sometimes means the loss of a crop of honey.

It is a mistake to attempt the production of honey, commercially, in a locality not suited for the business. The same may be said of queen rearing. It is a mistake to attempt it as a business in the Northern States—the seasons are too short. I followed it several years, and, while the experience may be valuable to me in my position, I am now well satisfied that I would have made vastly more money had I turned my whole time and attention to the production of honey.

It is a mistake to suppose that a poor location can be changed to a good one by planting for honey. Those who thus imagine do not realize the vast area of bloom that is needed to produce a surplus crop of honey. The bees of an apiary, going 2½ miles in every direction, scour a territory of about 12,000 acres. There is this to be said, however, if the soil, climate, and other conditions, are such that it is profitable for farmers to raise such crops as yield honey, then they will be raised, and the acreage will be such that the yield of honey from them will be of benefit to an apiary in that locality. Note the buckwheat regions of New York and the alfalfa fields of Colorado as examples.

The Influence of Locality.

IN my earlier bee-keeping years, I was often sorely puzzled at the diametrically opposite views often expressed by the different correspondents for the bee journals. In extenuation of that state of mind I may say that at that time I did not dream of the wonderful differences of locality in its relation to the management of bees. I saw, measured, weighed, compared, and considered all things apicultural by the standard of my own home—Genesee County, Michigan. It was not until I had seen the fields of New York white with buckwheat, admired the luxuriance of sweet clover growth in the suburbs of Chicago, followed for miles the great irrigating ditches of Colorado where they give life to the royal purple of the alfalfa bloom, and climbed mountains in California, pulling myself up by grasping the sage brush, that I fully realized the great amount of apicultural meaning stored up in that one little word—*locality.*

The basic principles of apiculture are the same the world over, but the management must be varied according to the locality. In the South and extreme West, the wintering of bees is easily accomplished; it being necessary, only, to see that they have sufficient food. As we go North, some protection must be given—either by packing or by the use of chaff hives. As we go still farther North, successful wintering is secured, as a rule, only by the use of first-class winter-stores, and putting the bees into a cellar.

In Cuba and Florida the honey harvest comes in the cooler part of the year, or what corresponds to our Northern winter, and those varieties of bees that will breed late in the summer, even though little or no honey is coming in, are more desirable; as more populous colonies are thus secured at the opening of winter. In the Northern States, east of the Mississippi, the main honey-flow comes,

The Author of Advanced Bee Culture Admiring a Luxuriant Growth of Sweet Clover.

as a rule, early in the summer. It may be very abundant, but is seldom of long duration; for this reason those varieties of bees are preferable that rear brood very abundantly early in the season, and then slacken breeding as soon as the main harvest begins. In some parts of the West the honey harvest is much longer than in the East. There are no such rapid flows as we have here sometimes from basswood, but there is a steady flow that may last for months; the conditions being ideal for the production of comb honey, as there is abundant time in which to build combs for the storage of the honey, fill them and seal them over.

In the white clover and basswood regions, swarming and the main honey-harvest come at the same time; in some parts of the Southwest, swarming comes on with the flow from the early, minor honey plants, and is almost entirely abandoned with the advent of the heavy honey-flow that comes on later.

The question of large versus small hives, over which there have been so many spirited discussions, is largely one of locality. In the cooler regions, where the harvest is early and short, small hives find favor, especially in comb honey production, while the large hive is a favorite in the warmer regions that are blessed with a long honey-flow.

Which the bee-keeper shall produce, comb, or extracted honey, is also largely a question of locality. Where the main honey-flow is short, as it often is from basswood, sometimes lasting only a few days, there is not time for the bees to build combs in the sections, fill them, and cap them over, before the harvest is over and past. With full sets of drawn combs in the extracting-supers, a good crop of extracted honey may be secured within a week. Such conditions as these exist in many parts of Wisconsin. Where honey must be shipped long distances to market, as is the case in Cuba and California, one very important reason for producing extracted honey is that there is so much less danger of damage in shipment. Dark honey is, as a rule, much more salable in the extracted form. When the flow is light but constant, and of long duration, as in Colorado, and the honey is white, comb honey production has its advantages, as honey is worth more when stored in sections than when taken in the extracted form.

California furnishes the most immense crops of honey that are any where produced, but they are entirely dependent upon the rainfall that comes in the winter. If the rains fail to come, the bee-keeper knows to a certainty that, not only will there be no surplus, but, unless the proper management is given, his colonies will perish from starvation.

A Field of Buckwheat in Full Bloom.

In the buckwheat regions of New York, not much dependence is placed upon the early honey-flows for securing a surplus. They enable the bees to breed up, and, as a rule, finish their swarming, before the buckwheat opens, when the main crop of the season is gathered. A colony so weak in the spring that it would be nearly useless in a flow from clover or basswood, has abundant time in which to build up for the buckwheat honey-harvest.

Then, again, there are localities near swamps, where the main flow comes very late, from fall-flowers, asters, and the like. The yield is often very abundant, but the quality is undesirable when used for winter-stores. If the cold confines the bees for several months upon such stores, they are almost certain to perish. The only remedy is to extract the honey and feed sugar syrup; unless it might be that of brimstoning the bees in the fall, and buying more in the spring from some other locality, a course which has been followed successfully, as the long season for preparation allows of the building up of one colony into several.

It would be an easy matter to use pages in giving illustrations of the differences in localities, but it is unnecessary; the thing for the bee-keeper to remember is that if he changes his locality he must leave behind him many of his old notions and methods, and seek the advice of his new neighbors who have been successful. The veteran bee-keeper from the verdant hills of old Vermont would make a flat failure were he to bring his apiary to Colorado, and manage it the same as he has been accustomed to doing. A bee-keeper can not know his locality too thoroughly. Some men succeed in localities where the majority fail, and one reason is because their more thorough knowledge of the locality enables them to adopt methods more perfectly adapted to the peculiarities of that location. Above all things, *know your locality*.

Best Stock and How to Secure It.

THERE are only two varieties of bees worthy of consideration for use in the United States; in fact, they are about the only varieties now left here for consideration, and they are the Italians and the Germans, or blacks, as they are commonly called. The prolific Syrians and the fierce, irritable Cyprians, have practically passed away on this side of the waters. These varieties of bees are very prolific, but undue prolificness is of no value—it is really objectionable for this part of the world. If queens cost large sums of money, there might be a shade of reason in desiring those that are prolific; but, to the practical honey producer, they cost almost nothing; and by using hives that are not too large, queens of ordinary prolificness will keep the combs sufficiently filled with brood. The great ambition of these varieties seems to be to rear brood, instead of to store honey. Their only object in gathering honey appears to be that it may be used in rearing brood. They will rear brood until the last drop of honey in the hive is used. The Syrians also have the undesirable trait of filling the cells so full of honey, and capping it so poorly, as to give it a dark, watery appearance, which is very objectionable in comb honey production.

Carniolans resemble the Syrians and Cyprians, so far as prolificness is concerned, but are very gentle, and cap their honey with a whiteness equal to that capped by the blacks, but this disposition to expend their energies in breeding and swarming, has caused them to be discarded in their purity, although a few bee-keepers still prefer a cross between them and the Italians.

In this matter of brood rearing, the Italians are unexcelled. During the spring months they push breeding with wonderful rapidity; but, as soon as the main honey harvest begins in earnest, breeding is greatly reduced, and most of the energy turned to the gathering and storing of honey. It might be safely said that the Italians are the standard variety of this country. They are very gentle in disposition, remaining quietly on the combs when being handled, while there seems to be about them a peculiarly quiet, steady, energetic determination possessed by no other variety. Almost any variety of bees will do fair work gathering honey when it is plentiful and near by, but when the flowers yield sparingly, and must be sought for far and wide, then it is that the Italians carry off the palm. For the production of extracted honey, the Italians are probably unexcelled, but in producing comb honey the blacks show two points of superiority. They are more willing to store their honey in the supers at some distance from the brood, and, in capping their honey, they leave a small space between the honey and the capping, which gives to the comb an almost snowy whiteness. The blacks are also more easily driven out of the supers with smoke, and more readily shaken from the combs. They are very irritable while being handled, many taking wing, and others running about upon the combs, gathering in bunches and dropping off upon the ground. For the production of comb honey there is probably no better bee than a cross between the Italians and the blacks, at least, so far as results are concerned. They are energetic workers, willing and ready to store their honey in the supers, but, unfortunately, they are possessed of a very uneven temper. Either variety, black or Italian, in its purity, is easier to handle than is a cross between them.

Modern bee culture, with its "bait" sections of partly drawn combs, or the putting on of extracting supers at the opening of the season, then changing them for sections after a start has been made, has well-nigh overcome the objection of the Italians clinging to the brood nest, while much can be done by selection in breeding to overcome the trait of poor capping. In brief, if I were to engage in the production of either comb or extracted honey, I should adopt pure Italians; then, by selection in breeding, get rid of the undesirable traits, such as "watery" capping of the honey, inclination to build large quantities of brace-combs, undue swarming, etc. Every bee-keeper of experience, who has tried different strains of bees, knows that there is a great difference between different strains of even the same variety. A bee-keeper who is just starting in the business, or one already in the business who has not taken such a course, ought to get queens from several of the best breeders, then adopt some easily kept but comprehensive system of recording the traits and peculiarities of each colony. The card system which has been so successfully adopted in so many ways, readily lends itself to this use. If the bees of any colony prove vindictive, re-queen it. If the bees of another colony are poor comb builders, or cap their honey poorly, destroy the queen and give them another. Do the same if they build large quantities of "brace-combs," or if they are unduly given to swarming, or if they are poor honey gatherers, or do not winter well. On the other hand, the desirable traits should be watched for and recorded, and queens reared from the queens of such colonies. Care ought also to be taken that no drones are reared, or allowed to fly, from undesirable stock, and pains taken to rear them in goodly numbers from the best stocks in the apiary. By pursuing this course, the bee-keeper will eventually build up a strain of bees that will be peaceable, hardy, good honey gatherers, and good comb builders. Well-directed efforts at improving his stock, carefully watching and recording the traits of each colony, getting rid of poor queens and keeping the best, perhaps buying queens occasionally and comparing their progeny with the stock already on hand, always breeding from the best, such a course as this will prove the most profitable of any which a bee-keeper can pursue. The wonder is that it is so greatly neglected.

The Choice of a Hive.

EARLY in every bee-keeper's life must come the choice of a hive—and a perplexing question it often proves. Probably there is no "best hive" for all persons, locations and uses; in fact, a choice is usually more or less of a compromise; the relinquishing of certain advantages for the sake of securing others considered more desirable. The tastes of a bee-keeper, his system of management, the kind of honey produced, the method of wintering, the location, etc., all have a bearing upon the kind of hive that is most desirable; but the inducements must be great, indeed, that would lead a man to adopt an *odd sized* hive or frame. As to size of frame, it is probable that the Langstroth is the most widely used, is well-adapted to the production of both comb and extracted honey, and its choice cannot be a serious mistake. With the choice of a frame, a decided step has been taken towards the choice of a hive; in fact, the most important question left to be settled, is the number of frames to be used in the hive. In those parts of the country blessed with a long honey-flow, or if extracted honey is to be produced, hives holding 10 Langstroth frames are desirable. If bees are to be managed in out-apiaries, or upon any plan where they are not to receive close and constant attention, large hives possess the advantage of containing sufficient stores to avoid danger from the bees starving in times of scarcity. The argument sometimes used in favor of large hives, that they give the queens more room to lay, is decidedly fallacious. We do not keep queens simply to "give them a chance to lay," but to secure the prompt and thorough filling of the brood-combs with eggs, and this is more surely accomplished by using a hive of moderate size, one below rather than above the laying capacity of the average queen. It is true that larger yields per colony may be secured with large hives, but not any larger yields

per *comb*. Where the honey-flow is short, or comb honey is produced, a smaller hive, one holding only eight Langstroth combs, has
its advantages.

The hive body for holding the frames need be nothing more than
simply a box, with plain, square corners, without top or bottom,
having rabbets on the upper, inner edges of the end-pieces, for supporting the frames. If a hive is nailed up with the heart side of the
lumber out, it is less inclined to warp. A plain, simple board,
cleated at each end, upon the under side, to prevent warping, with
half-inch strips nailed along the two sides and back end, upon its
upper surface, to support the hive, is the equal of any bottom-board.

A "Dirt Cheap" Bottom-Board.

It may not be amiss to say that hives may be used with *no bottom-
board* except the *earth*. The hive sits upon a rim made of rough,
cheap lumber, an entrance being furnished by making the front, end-
piece of the rim an inch narrower, and the rim filled with earth or
sawdust to within an inch of the top. At first thought this seems
like a very rough, primitive affair, as though using simply the earth
for a bottom-board would not answer, but it is difficult to say *why*.

A cover after the same style, simply a plain board cleated at the
ends to prevent warping, is a model of simplicity and desirability.
If it is difficult to obtain boards wide enough for covers, they may be
pieced, even made of narrow strips, then the upper surface covered

with a piece of muslin while the paint is fresh, and another coat of paint applied over the muslin. If kept properly painted, such covers will not leak.

Much has been written about staples and projections on frames to make them self-spacing, but the objections greatly overshadow the advantages, which are that all the frames are kept at *exactly* the same distance apart; in closing up the hive, or rather, in arranging the frames preparatory to closing the hive, they can all be shoved over in a body, by pressing against the outside one, and if the hives are to be moved, as from one apiary to another, the frames are held in position without any additional fastening. The moving of colonies from one location to another is something that occurs only occasionally, in many cases not at all, and it is better to specially fasten all of the frames once, or even twice, a year, should it become necessary, than to be continually annoyed by objectionable attachments. In closing up a hive there is some advantage in being able to shove the frames over without taking up any time in spacing them, but, so far as exactness is concerned, there is no necessity for self-spacing devices; as the combs may vary from 1¼ to 1½ inches from center to center, without any serious results. The eye and hand very soon become trained to sufficient exactness in the matter of spacing. The most serious objection to self-spacing is that it destroys the most valuable feature of hanging frames—the *lateral* movement. If frames hang free, it is an easy matter to press one over one way, and another the other way, and then lift out the one that hung between them. Self-spacing prevents this. Before self-spaced frames can be moved, a division board, or "dummy," must be pulled out at one side of the hive, and sometimes this board is glued fast and more difficult to remove than would be an ordinary comb. Another objection to staples, or other metal attachments, is that the honey knife is likely to strike them, and be dulled, when the honey is being uncapped; and they also give trouble by catching in the wire cloth forming the reel of the extractor. A few men have tried and been pleased with the plan of supporting frames upon nails driven into the centers of the ends of the top-bars. To illustrate: Take an ordinary Langstroth frame, saw off the projecting ends of the top-bar, then, into the center of each end of the top-bar drive a six-penny nail at such a point that its lower side will occupy exactly the same point as the lower side of the wooden projection occupied before it was sawed off. The nail is, of course, driven in until it projects exactly as far as the former wooden projection extended beyond the end-bar. These nail-supporters may be used either with or without metal rabbets. In either case the points of contact are so slight that there is little opportunity for gluing them fast, and the frames can always be loosened with the fingers.

Closed-end frames, in common with other styles of self-spaced frames, possess the advantage that they need no fastening when the hives are moved from one part of the country to another, but, aside from this, the advantages are all with the loose, or hanging, free-swinging frame.

A divisible-brood-chamber hive, one having two sets of shallow frames, thus allowing the hive to be divided horizontally, possesses some advantages. For instance, at the beginning of the season it is desirable to induce the bees to spread out and fill their combs as completely as possible with brood, and by dividing the brood nest horizontally, transposing the sections, placing the lower one above and the upper one below, we bring together, in the center of the hive, the outside, or spherical portions of the brood-nest, while the broad, center-surfaces are thrown to the outside. In their efforts to bring the brood-nest back to a spherical shape, the bees remove the honey from the center of the hive and replace it with brood, thereby increasing the amount of the latter. The transposition of the two sections of the brood-nest also throws a large surface of brood up close to the supers which greatly hastens the beginning of work in the sections.

The use of this style of hive also allows of contraction of the the brood nest without reducing the supering surface, or the bringing in of "dummies," as must be done with other styles of hive. Divisible-brood-chamber hives cost considerably more than other styles of hives, and, after using them for years by the side of the ordinary Langstroth hive, seeing them used by other persons in different locations, and considering the new features that have recently sprung up in bee-keeping, I have gradually come to the decision that if I were now starting in the bee business, I should not use the horizontally-divisible hive. In my opinion, its greatest point of superiority is in practicing contraction of the brood-nest; but so far as handling frames is concerned, there is no frame that approaches the plain, all-wood, hanging frame, and, in managing out-apiaries, in which case there is not time for using the bee-escape, this is a most decided advantage.

In northern climates, bees need more protection in winter than is afforded by a single-wall hive. In Michigan this is best afforded by a cellar; further south, some kind of packing is probably preferable. Whether this packing shall be in the shape of the so-called chaff hive, or in something of a temporary nature that can be removed in summer, is a point upon which bee-keepers differ. It is true that temporary packing calls for extra labor (but it does not come at a hurrying time of the year), and there *was* a time when it also resulted in some untidiness and unsightliness in the apiary during the winter, but the neat outer case and improved methods of packing that are now being adopted, have removed the latter objection, and greatly reduced the former. These methods of temporary packing are cheaper than the chaff hives, while the advantage of having light, single-walled hives during the working season, hives that can be picked up, handled, manipulated, tiered-up, carried, if advisable, to a distant or more desirable location—hives, in short, that can be handled in a way that means business—all these advantages are so great that I should never think of adopting the chaff hive. I know there are methods of management in which the unwieldy, stand-still character of the chaff hive proves no obstacle; but such methods are not the most expeditious.

In brief, my choice of a hive for Michigan is a simple, plain box with plain, all-wood hanging frames—and I would winter the bees in the cellar.

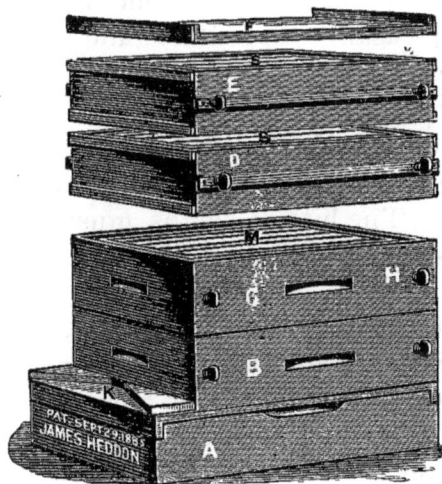

DIVISIBLE-BROOD-CHAMBER HIVE.

Honey Boards and Queen Excluders.

WITH the majority of frames in use, bees build little bits of combs between the top bars of the frames, and, extending the combs upwards, connect them with the cover of the hive, or the bottom of a case of sections, or whatever is next above the tops of the frames. These little bits of combs are called brace combs or burr combs. It is very unpleasant, unprofitable and untidy to lift off a case of sections, and, in so doing, pull apart a net work of combs that connect the bottoms of the sections with the tops of the brood frames. The honey drips and daubs about and attracts robbers, if there are any to be attracted. The bits of combs must be scraped from the bottoms of the sections, and the muss cleaned up as best it may.

The bee-keeping fraternity is, I believe, indebted to Mr. James Heddon for the modern honey board, which practically does away with all of this trouble from brace combs. This honey board is simply a series of slats fastened to a frame as large as the top of the hive, and placed over the brood nest. These slats are about 5-16 of an inch thick, placed ⅜ of an inch apart, and of such width and so arranged that each opening between them comes exactly over the center of the top bar of a brood frame below. In other words, the slats break joints with the top bars of the frames below.

As the tops of the frames are ⅜ of an inch below the level of the top of the hive, there is a ⅜-inch space between the tops of the frames and the bottom of the honey board. The outside rim or frame-work of the honey board is ⅜ of an inch thicker than the slats, thus the surplus case is raised ⅜ of an inch above the slats of the honey board. In short, the honey board is a series of slats, ⅜ of an inch apart, placed between the brood nest and the supers, with a "bee space" both above and below the slats. In the space below, between the slats and the brood nest, the bees build brace combs *just the same as ever*, but, for *some* reason, the space *above* is almost always left free from the disagreeable brace combs. A case of sections can be lifted off as clean and free from daub as when first placed upon the hive.

There have been more or less successful attempts to do away with the necessity for a honey board by using wide, deep top bars, *accurately spaced*; and while such an arrangement does away with a large share of the burr comb nuisance, I have yet to see a case in which there was not enough of it left to warrant the use of a honey board.

A slatted honey board is easily made queen excluding by simply cutting saw kerfs in the edges of the slats, and slipping strips of perforated metal into the kerfs between the slats. Whole sheets of zinc have been used as honey boards, but they are lacking in rigidity. They are likely to sag, bend or kink, thus de-

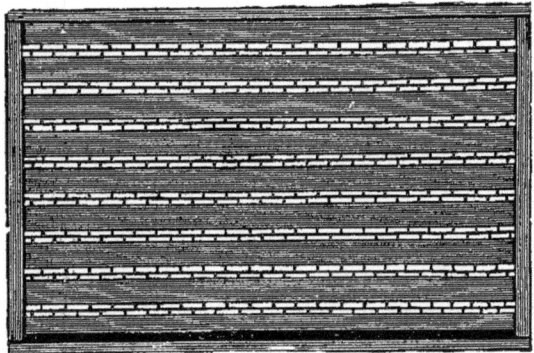

Queen Excluding Honey Board.

stroying the perfection of the bee spaces. If a sheet sags, the space above becomes so large that there is a likelihood of comb being built therein; while the space below becomes so small that propolis is placed between the zinc and the tops of the brood frames. The wood-zinc honey board is free from this defect.

In the production of comb honey there is little need for a queen excluder over an old, established colony; but when a swarm is hived in a contracted brood chamber having starters only in the frames, and given the supers of partly finished sections from the old hive, a queen excluder is almost a necessity. The queen, finding no combs in the brood nest, at once invades the sections, where the bees soon clear out some of the cells for her to lay in, and, having begun her brood nest there, she is quite likely to remain there until considerable comb has been built below.

In the production of extracted honey, queen excluders are a great convenience, if not a necessity. If they are not used, the brood is almost certain to be scattered through the supers, or upper stories; and ripe honey, ripe as it ought to be when it is extracted, cannot be thrown from the combs very rapidly or completely, without at the same time throwing out the brood. If brood is found in the upper story, it is, of course, sometimes possible to exchange such combs for the outside combs in the brood nest, if such can be found without brood, but all this takes time. To successfully conduct an apiary, the fixtures and methods should be such that the work will move along smoothly, and in a systematic manner, without any "hitches."

There is also another point to be considered in connection with the use of queen excluders when producing extracted honey, and that is the freeing of the supers of bees by the use of bee escapes. If the super contains brood and, perhaps, the queen, the bees could not be induced to desert by the use of an escape. If they did leave the brood, then something would have to be done with the brood, as already mentioned. In short, advanced bee culture has divided the hive into two distinct apartments—brood and surplus—and unless this division can be maintained, many profitable plans must be relinquished. The queen excluding honey board enables the beekeeper to thus set a boundary, beyond which the brood can not go.

Sections and Their Adjustment on the Hive.

FOR making sections, basswood is used to a greater extent than any other wood. It is the whitest, readily obtainable in many parts of the country, while it possesses the elasticity needed in the one-piece section. Its faults are that it shrinks and swells badly, becomes mildewed and discolored very easily, and any honey dropped upon it soaks in and leaves a stain. White poplar is the best wood for sections. It is whiter than basswood, very hard, does not shrink or swell readily, and is not stained by contact with honey, or easily soiled by handling; but it lacks the elasticity necessary in the one-piece section. There are no handsomer nor better sections made than the four-piece, white poplar, and the only valid objections that can be brought against them are that they cost more and that more time is required in putting them together. I am aware that I am always pleading for time-saving fixtures, but there must be a distinction made between the hurry and bustle of swarming-time and the leisure of a winter's evening; or between the time of an experienced apiarist and that of some boy or girl who can put together sections. The objections to the one-piece sections are that they can not be made of the most desirable wood; that, as usually made, they do not remain square when folded; and that they are made with "naughty" corners which gouge into the honey when crating it or removing it from the crate. When separators are used, the latter objection is removed. The reason why the so-called "naughty" corner is always found upon the one-piece, bee-way section is because the openings in the top and bottom bars can not be cut clear through to the side bars, as the small film of wood left to hold top and side bar together is then more likely to break. When the opening extends clear across, as in the four-piece section, the combs are more completely built out and attached to the top and bottom bars. The top and bottom bars of sections ought to be ⅜ of an inch narrower than the side-bars. Usually, the top and bottom bars of sections are too wide, leaving too narrow openings between them. To put the matter in a few words, the one-piece, although possessing some faults, is cheaper and can be put together quicker than the four-piece, which costs more, but is *faultless.*

There is also another point that is coming rapidly to the front, and that is the scarcity of basswood timber. Possibly the same may be true of white poplar, but there are other white woods, hard maple, for instance, from which four-piece sections can be made.

The standard size of sections is 4¼ x 4¼ inches; at least, this has been the standard for many years, and I think is yet, but there is considerable effort to place upon the market, and secure the adoption, of a tall section—about 4 x 5 inches. Its chief advantage is in being more pleasing to the eye—possibly in conveying the impression that it contains more honey than a square form having the same amount of surface. Our windows, our books, our pictures, etc., are made oblong instead of square, because they are more pleasing to the eye, and, for the same reason, a tall section presents a more pleasing appearance than a square one, but I do not consider this point of sufficient importance to warrant a bee-keeper in changing his fixtures in order that he may use the tall section.

Thus far, in this chapter, we have been considering what are called "bee-way" sections, those in which the bees gain access to them through insets, or "bee-ways," cut in the top and bottom bars of the sections, but, of late, there has been introduced a new style, called the "plain" section, in which there are no insets, it being the same width all the way around. Sections of this style are held bee-space apart, and the bees admitted, by the use of what are termed "fence" separators, from their resemblance to a board fence. A fence separator is formed of slats about ⅛ of an inch in thickness, held a bee-space apart by cleats glued, or nailed, in an upright position to their sides. These cleats, or posts, are of such a thickness, and placed at such a distance apart, upon each side of the separator, that the edges of the side bars to the sections come against them, and are thus held out bee-space from the slats.

The principal advantage of plain sections and fence-separators is that the freer communication thus allowed the bees induces them, from some reason, to build out the combs fuller around the edges and corners, and attach them more perfectly to the sections. This

Plain and Old-Style Sections—Plain Sections in the Lower Row.

gives the finished product a more attractive appearance, and greatly lessens the danger of breakage in shipment. Another point, although it may be a minor one, is that a plain section is filled fuller of honey; that is, the edges of the wood do not stand out so far above the surface of the comb as they do in the bee-way sections. A filled plain section has a plumper look than a bee-way section, the latter having the appearance of being only partly filled. A tall, plain section may not contain any more honey than a square section of the bee-way type, but it *appears* to contain more, and has, withal, a more attractive appearance. There is still another little point, and that is that a plain section offers special advantages in the matter of cleaning it of propolis, as there is no inset to work into with the scraping knife. I do not, however, consider the advantages of the plain section sufficient to warrant any expensive change of fixtures in order that it may be adopted.

While I have produced tons of comb honey without the use of separators, and could do it again in this locality, I think I should use them were I again to engage in comb honey production; I know of no objection to their use, except that of cost, and I certainly would advise their use by the great mass of bee-keepers. In many localities there are bee-keepers who can, without separators, produce sections of honey that are tolerably perfect, straight enough to be crated with a little care, but there is another end to the business — that of the retailer and his clumsy clerks who are not bee-keepers. Nothing discourages and disgusts a retailer more than a lot of dauby, dripping, damaged sections.

Perhaps I am a trifle old-fashioned in some respects, and one is that if I were to adopt the old-style, bee-way section, I should also adopt wide frames and tin separators. I may be notional, but the so-called section holders (wide frames minus top bars) seem like an incomplete affair to me. When a wide frame is used the sections are protected on all sides, and come off the hive in all their virgin whiteness. By the addition of a top bar, thus making a wide *frame*, a tin separator can be used, when fussing with separators (by their breaking and splitting) is done with for all time. To my mind, wide frames with tin separators furnish the most perfect method of adjusting bee-way sections on the hive; and, before closing, let me tell how to put the tins on in such a manner that they remain taut. Nail two blocks of wood upon the top of the work bench at such a distance apart that the top and bottom bars of a wide frame (after it has been put together) can just be "sprung" in between the blocks. This shortens the distance between the end bars. While held in this position, nail on the separator. Upon removing the frame from

ADVANCED BEE CULTURE. 45

between the blocks, the top and bottom bars will straighten out, and in so doing, draw the tin as taut as a drum head.

If a man *can* succeed, to his satisfaction, in producing comb honey without separators, then I know of no more desirable super than the old-style Heddon. This is a box, without top or bottom, the size of the top of the hive, and a bee-space taller than the height of the sections, having upright, wooden divisions, as wide as the sections are tall, put in crosswise of the case and at such a distance apart that a row of sections will just nicely slip down between. Flat against the lower edge of each division is nailed a strip of tin 1-4 inch wider than the partition is thick. The edges of the tin, projecting out ⅛ of an inch, on each side, beyond the sides of the division boards, afford a support for the sections. I have used hundreds of these supers for years; in fact, produced nearly all of my comb honey in them, and, if a man prefers the old style, bee-way sections, and does not care to use separators, this super is simply perfection.

To sum up this chapter in a few words, my preference is for a tall, plain, four-piece section of white poplar, used with fence separators.

Arrangement of Hives and Buildings.

IN a small apiary, the matter of arrangement is not of great importance, but as the number of colonies begins to approach 100, the question of arrangement becomes one of considerable importance. Two things need consideration: the convenience of the operator, and the giving of such an individuality to each hive that each bee can readily distinguish its home.

Before discussing these points, it might be well to say a few words about the location of the apiary. First, it ought to be some distance from the highway. What that distance should be, depends upon what there is between the bees and the street. If there are buildings or trees, or even a *high* fence, the bees may be quite near the road; as, in their flight, they rise above these obstructions, and thus fly over the heads of the passersby. If there is nothing between the apiary and the highway, the apiary ought not to be nearer the street than ten rods, and fifteen or twenty rods would be better. It is possible with a small apiary to avoid trouble even if it is located near the street. When it is necessary to handle the bees when no honey is coming in, and such handling is likely to irritate them, the work can be done just before dark, when the bees will not fly far from their hives; but in a large apiary there is too much work that must be done when the bees may not be in an amiable mood, to enable the operator to perform it during the twilight of evening. If necessary, the bee-keeper can protect himself with a veil, and, armed with a smoker, he can go on with the work, even if the bees are a little "cross," but the apiary must be isolated.

Nearly level ground is preferable in an apiary. If it slopes gently to the south or east, so much the better. It should never be in such a location that the water will stand upon the ground.

I have tried placing the honey house in the center of the apiary, and having the hives in long double rows that radiated from the honey house as the spokes in a wheel radiate from the hub. In each double row a space large enough for a wheel barrow is left between the hives, and the entrances of the hives are turned away from the path left for the operator and his wheel barrow. So far as reducing the labor of going to and from the hives is concerned, this arrangement is excellent, but it has the quite serious objection that only part of the apiary can be seen at one glance from the honey house. In watching for swarms it is necessary to look in *four* different directions in order to ascertain if a swarm is out. When the house is at one side of the yard, the whole apiary can be taken in at a glance. Other things being equal, the south side of the apiary is preferable for the house. In looking for swarms the bee-keeper does not look towards the sun, but has the clear northern sky for a back ground, while the shady side of the building, which will be naturally sought by the tired bee-keeper as the best spot in which to take a breathing spell, is towards the apiary.

Most bee-keepers are in favor of having the building two stories high, using the upper story as a store-room for hives and fixtures, the lower story for work shop and honey room, the latter being partitioned off by itself, and the cellar under the building for wintering the bees. The usual mistake in making such buildings is in not having them large enough. The honey room ought to be located in a southern corner of the building, and the walls made of some non-conductor of heat. Some even paint the side of the building a dark color where it comes over the honey room, in order that as much as possible of the sun's heat may be absorbed. The idea is that the honey must be kept as warm as possible. If there is any unsealed or unripe honey, this high temperature causes evaporation and improvement. By keeping such a room warm with a stove in winter, comb honey has been kept over until another year, and actually improved by the keeping.

But, to return to the arrangement of hives. When the honey house is at one side of the apiary, the hives may still be arranged upon the radiating plan, by having the rows radiate from the honey house door, thus forming one-half of a large wheel, instead of the whole of a smaller one, as in the case of having the honey house in the center. When the radiating rows are very long, they become far apart at the outer ends, or else very close together at the inner ends. To remedy this, shorter rows, or "spurs," may be put in between the long rows at their outer ends.

Another arrangement is that of placing the hives in a hexagonal manner, each hive being the center of six others. This is a pleasing arrangement to the eye, but it has been reported that the massing of the hives in such a regular manner has a tendency to lead the bees to enter the hives standing on the outside of, or edge of, the apiary, thereby weakening the colonies in the center of the yard.

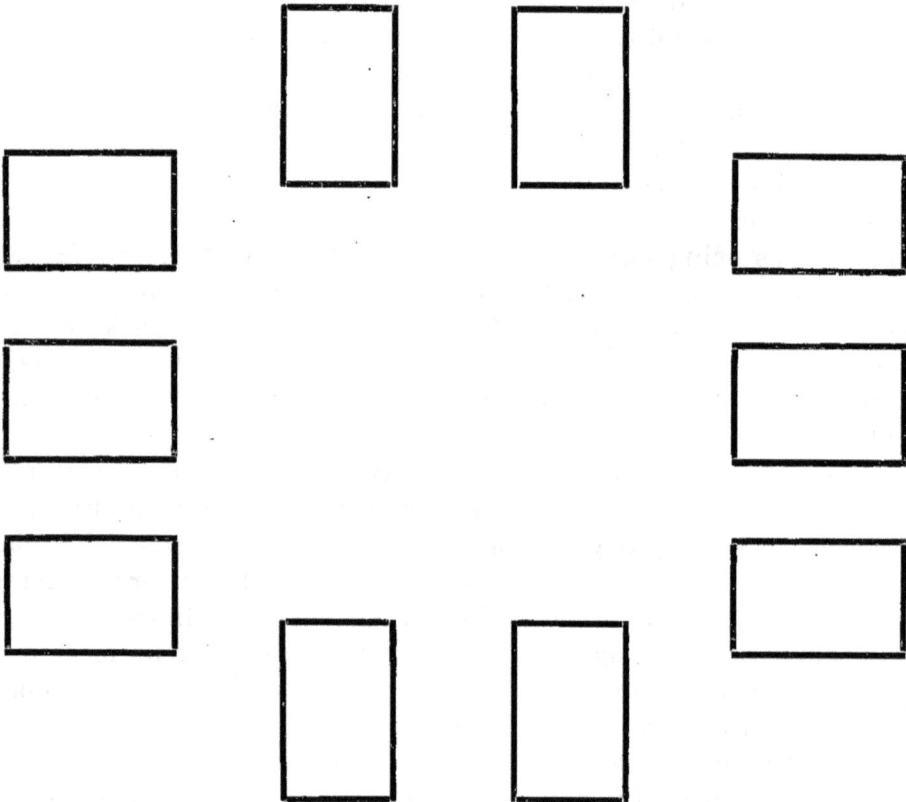

Placing the hives in small groups is a most excellent arrangement. Mr. J. E. Crane of Middlebury, Vermont, arranges his hives in groups of ten each, each group being arranged as follows: Two hives facing the north, three facing east, two the south, and three the west. Nine such groups, arranged in a square, three groups each way, furnish room for ninety hives in a very compact body, yet each hive is given a most distinct individuality.

Still another arrangement is that of placing the hives in circles. The entrances of the hives in the inner row are towards the center, while those of the outer row face outwards. This leaves the space between the two rows comparatively free from bees, and, the operator can work in this space without annoyance to himself or the flying bees. If the two circles do not furnish sufficient room, more and larger circles may be added, or there may be two sets of circles, or three sets, arranged in the form of a triangle, or even four sets arranged in a quadrangle.

In nearly all of the large apiaries that I have visited, the hives were arranged in straight, simple rows, like the squares of a checker board, the entrances in some instances, facing the same way, while hives were from six to eight feet apart. I would prefer to have the entrances of each alternate row turned towards the ·east, and the entrances of the hives in the other rows turned towards the west. This would leave each alternate passageway comparatively free from bees, and the operator could work here without the bees bumping their heads against his. I would prefer to have the entrances to every hive face either east or west, because I wish to shade each hive with a light board, 2 x 3 feet in size, laid over each hive, and projecting towards the south, and this projecting board would be in the way of the flying bees if the entrance were upon the south side. When the hives are arranged in rows radiating from a common center, I always turn the entrance of each hive so that it is either east or west.

There is no reason for placing hives farther apart than is necessary to afford sufficient space on all sides for the operator. Bees do not locate their hive so much by the distance that it is from other hives, as they do by the surroundings; and these surroundings are usually other hives. To illustrate: Let the end hive be removed from a long row of hives, and the bees belonging to the removed hive will almost unhesitatingly enter the hive that has *become* the end hive in the row. Two hives may stand side by side, perhaps almost or quite touching each other, yet each bee has no difficulty in distinguishing its own hive. In a row of three, or four, or even five hives, the same might be said, but, as the number goes beyond this, there is a little uncertainty about the matter. When their hives are in long rows, some bee-keepers arrange them in groups of three or five in the row, leaving a wider space between the groups than there is between the individual hives composing a group.

The greatest objection to any uniformity of arrangement that makes it difficult for the bees to "mark" their location, is that queens may enter the wrong hive upon their return from their "wedding trip." With my method of management, in which the hive with a young queen is given a new stand to prevent after-swarming, *a la* Heddon, this difficulty is easily remedied by placing the hive in some location that is easily marked—the end of a row, for instance. When this cannot be done, the hives containing unfertile queens may be

marked in some conspicuous manner that will easily enable them to
distinguish their own hives.

.In queen rearing it is important that the small hives, containing
the nuclei, be scattered about promiscuously; the greater the irregularity and oddity of the arrangement, the less will be the loss of queens
from their entering wrong hives; but, in a large apiary managed for
honey, it is doubtful if there is a more practical arrangement than
that of placing the hives in rows; and it seems to me that a little is
gained, and nothing lost, by having the rows radiate from the honey
house door.

Comforts and Conveniences in the Apiary.

B Y these are meant those things not absolutely essential to success, but that serve to render more smooth and pleasant the somewhat "thorny" path trodden by the bee-keeper. To illustrate: Mr. H. R. Boardman, of Ohio, has a cart, for carrying his bees to and from his bee cellar, with which there is no necessity for even lifting the hives to place them on the cart. It is made like a wheel barrow with two wheels, and having two long prongs projecting in front. When the cart is wheeled up to a hive, one prong goes one side of the hive and the other the other side, when, by depressing the handles, the hive is lifted from the ground; cleats upon the sides of the hive prevent it from slipping down between the projecting prongs. Then, again, Mr. J. A. Green, of Colorado, has an arrangement for opening the honey house door by simply stepping upon a pedal. When both hands are occupied with tools, a case of honey, or something of that sort, such an arrangement is quite a comfort. Mr. Green is also the man who keeps kerosene oil in a spring-bottom oil can to squirt on the fuel in a smoker when "firing up."

Most of these comforts are comparatively inexpensive. To think of them and secure them is often more work than to earn the money with which to buy them; but their possession often makes all of the difference between a season of pleasure and one bordering on drudgery, to say nothing of the bearing they may have upon the profits. These little helps and conveniences are, in one sense, the oil that makes the great apicultural machine move smoothly, and I believe it worth while to enumerate a few of them.

I will begin with the bee-keeper himself, or rather with his clothing, as his comfort is largely dependent upon that. When there is much shaking or brushing of bees to be done, I prefer to wear light, calf skin boots with the trousers tucked inside. If the grass is wet, I wear rubbers over the boots. When shoes are worn, the trousers may be tucked inside the stockings. One fundamental principle about clothing to be worn in the bee yard is that one garment laps *closely* over the other, leaving no opening into which a bee can crawl, and the lower garment should lap *over* the upper one, as bees almost invariably crawl *upwards*, and the clothing should be so arranged that a bee can crawl from a man's foot to his head without being led into any opening. Mr. Arthur C. Miller suggests canvas shoes that lace well up around the ankle, such as are worn by tennis and base ball players and cyclists. Then he would have the trousers come just below the knee, with canvas leggings to cover up the lower part of the legs. His ideal coat is a close-fitting jacket of light-weight that buttons up to the throat. In the heat of the day, however, few bee-keepers, doing active work in the apiary, need either coat or vest. The hat that approaches the nearest to perfection, in Mr. Miller's opinion, is the helmet. It has visors front and back, and a ventilator all around the rim and the inner band. It is light and cool, and protects both the eyes and the back of the neck from the sun. Such suits as those described by Mr. Miller can be had in white or colored duck, and are light, cheap, washable and serviceable; and complete, or in part, are worn by cyclists and others. Light colored clothing is not only cooler, but saves the wearer from some stings, as the bees seem to have a decided aversion to dark or black objects. I know one bee-keeper who dresses in white duck from head to toe, and he is positive that it saves him from many attacks from the bees. In the heat of the working season I wear linen trousers, a white cotton shirt and a straw hat. I have seen recommended the wearing of light woolen clothing, but have never tried it. Ernest Root mentions the comfort that he has derived from the wearing of light underclothing, part woolen. But he does not perspire freely, and his underclothing retains the perspiration, keeping the skin moist. With me it is the reverse. I perspire so freely the clothing is soon soaked through and through, and frequent changes are necessary. Perhaps each will be obliged to decide the matter by personal experience.

I never wish a veil attached to the edge of the hat rim. It is only part of the time that a veil is needed, and when it isn't needed I wish it off out of the way. I prefer a veil with a string run into a hem around the top, then the upper edge can be puckered up until it will just slip down nicely over the hat crown. And when it is necessary to wear a veil in hot weather, who has not wished that there was some way of holding it down, aside from that of tucking it inside the collar ? When the neck is hot and sweaty, how uncomfortable it feels with a sort of muffler pressed close against it by the collar. Besides this, the veil is held suffocatingly close to the face—so close, too, that the bees often sting through it. All this may be avoided, and I'll tell you how. In a hem in the bottom of the veil run a string, leaving about a foot of the hem, right in front, unoccupied by the string. That is, let the string enter the hem at about six inches to the right of the center of the front, pass it around the back of the neck, bringing it out of the hem at a point about six inches to the left of the center of the front. The projecting ends of the string must be long enough to pass under the arms, cross at the back, and then be brought around and tied in front. The string holds the edge of the veil securely out upon the shoulders, while, if the right length of hem is left without a string in front, that part will be drawn snugly across the breast. To Mr. Porter, of bee-escape fame, belongs the honor of devising this unsurpassable way of holding down a bee veil.

Gloves I have never worn, and doubt if I could be led to believe them a comfort. To use them would seem too much like a cat with mittens on trying to catch mice. Perhaps a beginner might tolerate them until his timidity had worn off.

I know of no comfort in the apiary greater than a smooth surface (of earth) thickly covered with grass. A lawn mower can scarcely be called a comfort—it is a necessity. Sprinkle salt around the hives to kill the grass a distance of six inches from each hive, then the lawn mower can cut all of the grass that grows.

About the first thing needed upon beginning work in the apiary is a smoker; and oh, how much comfort or discomfort can come through this little implement. If any of my readers have suffered from smokers that spill fire, that become stopped up with soot, that go out, or from fuel that will not burn, let them get a Bingham smoker, of the size called the "Doctor," get a barrel of planer shavings from dry pine, to use as a fuel, and then take comfort. If you have never used shavings as a fuel, you may have trouble in getting the fire to going. Don't put in too many shavings at first. Leave off the cover and keep puffing until they have burned down to cinders before putting in more. A little kerosene oil, from an oil can, as has been mentioned, is a great help in starting the fire. When through using the smoker, don't throw out the fire, but stop up the nozzle with a wad of grass, thus smothering the fire, and the charred remains left in the bottom of the smoker will light very readily at the touch of a blaze from a match—much more so than with fresh fuel. Keep matches in a safe place near where the smoker is to be lighted. Never be pestered by having to run off somewhere after a match. Above all, don't keep the smoker fuel and matches in the honey house; the danger from fire is too great. Rig up a box, or a barrel, or an old bee hive, with a rain proof cover, and have it located some distance from the honey house. I kept the fuel in an old wash-boiler, and had it "burn out" once. As it was out of doors, no harm was done. Keep the cap of the "Doctor" filled with green weeds or grass, and there is no danger of blowing sparks into the hives.

Have a wheel barrow for carrying cases, hives of honey, and other heavy articles. In making a wheel barrow, some bee-keepers have used a wheel from some old, discarded bicycle, the pneumatic tire doing away with the jolts in carrying honey or hives of bees.

With such hives as I use, the cover can be turned up on edge and used as a seat; where such is not the case, a seat of some kind ought to be provided. A light box 17 x 12 x 9, gives a chance for having a seat with any one of these heights. It should be strong enough not to rack, and have hand holes in the side for carrying it.

A hammock in the shade of a tree, or in the work shop, is a great comfort. Ten minutes rest in a reclining position are of nearly twice the value of that taken in a standing or sitting posture.

For brushing bees off the combs I know of nothing more effectual or comfortable to both the operator and the bees, than the so-called Coggshall brush, which is a sort of a whisk broom, with the strands thinned out, and longer than the ordinary whisk broom, so as to enable the operator to give a soft, pliable, easy sweep of the combs. In using this brush it is not intended that the combs be brushed with the *ends* of the strands, as one would sweep a floor; instead, the brush is laid flatwise against the comb, and given a quick, sharp, lateral sweep.

A close rival of the Coggshall brush is the Ferry, double brush. With it *both* sides of the comb may be brushed at the same time. The whole thing is so simple that the only wonder is that it was not thought of before this. First there is a long loop of spring-metal something like an ox-bow, or hair-pin. Upon the inner sides of the spring are fastened two opposing brushes of long bristles. A slight pressure upon the spring, as it is held in the hand, brings the brushes in contact with *both* sides of a comb, when a downward sweep or two will free the comb of bees. Another thing must not be overlooked (and it is of much importance), and that is that it has a chain attached, and to the end of the chain is fastened an eyelet that may be slipped over a button, thus enabling the operator to always know where his brush is without having to hunt for it.

Let each bee-keeper look about his apiary and see if he is not doing some of his work in an awkward manner, one that might be avoided by the providing of a few comforts and conveniences.

A Well Shaded Hive.

Shade for Bees.

SHALL we shade our bees? If so, why, when, how? Some bee-keepers do not shade their hives; others do. Why do they do it? Is it really necessary? Do they thereby secure more honey? These are pertinent questions to which it is difficult to give definite answers, but about which it is advisable to know all that *is* known.

The temperature of a colony of bees in summer, when brood is being reared, is nearly 100 degrees. Until the temperature in the sun reaches this point, shade is no benefit; rather it is an injury, as it deprives the bees of the warmth of the sun at a time when it would be of some benefit. When the temperature in the sun goes above 100 degrees, and begins to climb up to 110 degrees, 120 degrees, 130 degrees, *then* the effort on the part of the bees is to lower instead of raise the temperature of the hive. Crowds of bees stand at the entrance of the hive, and with their wings create strong, ventilating currents of air. It has been asserted that the bees leave the combs of honey well-nigh forsaken when the temperature is very high; the reason given being that the combs can be kept cooler when not covered with bees. I have also read and been told that bees would "hang out," that is, cluster upon the outside of the hive, instead of working, if their hives were left unshaded during a hot day; that they are compelled to thus desert their hives to save their combs from destruction. I have always kept my hives shaded during hot weather, hence cannot speak from experience upon this point; but, if it is true, then it would seem that shade, in very hot weather, is both desirable and profitable. This much I have noticed, that weak colonies, nuclei, for instance, seldom make any demonstration of discomfort from heat, even when left unshaded, while strong colonies are puffing and blowing like the runner of a foot-race. Why is this? Isn't it because the strong colony is suffering from the accumulation of its own heat—that generated by itself—that can not escape fast enough? If this is true, why isn't a chaff hive the most insufferably hot place imaginable for a colony of bees in hot weather? Possibly the point is just here: the bees in the chaff hive have to contend with their own heat only, while those in the single-wall hive have that from the sun in addition to their own. The thick walls act as a sort of absorbent of heat; taking it up during the day, and gradually giving it up during the cool of the night. Let this be as it may, a colony can be kept the coolest in a thin-wall hive in the shade. How do *we* keep cool in hot weather? We wear thin clothing, and lie in the hammock in the shade. A colony of bees is a living, heat-producing body, and can be kept cool in the same manner that we keep our bodies cool, viz., let its clothing (hive) be thin, with a free circulation of air upon all sides, above and below, and then protect it from the sun's rays.

The color of the hives has a great bearing upon the necessity for shade. Black, or a dark color, absorbs heat, while it is reflected or repelled by white. I have seen the combs melt down in an old, weather-beaten hive that stood in the sun, but I never saw them melt in hives painted white, even if standing in the sun.

There is still another point that has a bearing upon the question under discussion, and that is the circulation of air about the hives. I have read of combs melting down in hives standing in shade so dense that the sun never shone upon them. The trouble was that growing corn on one side, and dense brush upon the other, made it so close that no air circulated.

Shade is not needed in the spring, fall, morning or evening. The only time that it is needed, if it *is* needed, is the middle of our hottest days; and some temporary, quickly adjustable, easily removable shade is preferable to an attempt to furnish a permanent shade by growing evergreens, grape vines and the like. In fact, a permanent shade, like that furnished by an evergreen, is an injury in spring, robbing the bees of the benefit to be derived from the heat of the sun. In fact, I know of nothing better than a light board, 2 x 3 feet in size, laid upon the top of the hive. One of the longest edges of the board is placed parallel and even with the north edge of the top of the hive, the opposite edge of the board projecting beyond the hive. This shades the hive when shade is needed, and only when it is needed—in the middle of the day. In a windy situation it may be necessary to lay a brick or stone upon this board to keep it in place. Don't imagine that hooks or something of that kind will be preferable for holding the shade-boards in place. A weight is the simplest, cheapest and most convenient. I make these shade-boards by nailing the thick ends of shingles to a piece of inch board four inches wide and two feet long. They cost only five cents each, and, in the fall, they can be tacked together, forming packing boxes in which to pack the bees for wintering.

For the comfort of the apiarist, it is well to have a few scattering trees in the apiary, but let their branches be trimmed to such a height that they will not be knocking off his hat, or gouging out his eyes.

Perhaps this whole matter of shade might be summed up something as follows: If the apiary is located where the cool breezes can fan the heating sides of the hives, wafting away the heat ere it accumulates, and a broad, generous entrance is furnished each tidy, *white* hive, I am persuaded that shade is not so *very* essential; but, if the hives are dark in color, or the apiary located where there is not a free circulation of air, I feel sure that shade is an absolute necessity to prevent the combs from melting, if for nothing else.

ROYAL PALMS OF CUBA.

Contraction of the Brood Nest.

THE brood nest is contracted to prevent the production of brood at a time when the resulting bees would come upon the stage of action at a time when there would be no honey to gather—when they would be consumers instead of producers. It is also contracted to compel the bees to store the honey in the sections instead of in the brood nest. There are several reasons why this is desirable. The honey from clover and basswood is white, fine flavored, and brings a higher price than that gathered later; hence it is more profitable to force this higher priced honey into the sections, and allow the bees to fill the brood combs, later in the season, with winter stores from such sources as yield honey that brings a lower price. When it is desirable, either from its cheapness, or from its superiority as a winter food, to use sugar for winter stores, contraction of the brood nest can be so managed as to leave the bees almost destitute of honey in the fall, which does away with the trouble of extracting, and leaves nothing to be done except to feed the bees. Such, in brief, are the advantages of contracting the brood nest. Where the honey flow lasts nearly the whole season, with no long periods of scarcity, and the quality of the honey is uniform throughout the season, and no advantage is to be found in substituting sugar for honey as winter stores, I see little need of contracting the brood nest, and would advise that it be of such size that an ordinarily prolific queen can keep the combs well-filled with brood in the early part of the season; but where any of the first mentioned conditions exist, the bee-keeper who neglects "contraction," is not employing all of the advantages that are available.

As a rule I don't advise the contraction of the brood nest of an established colony. If it does not properly fill its hive, is too weak, and the time for putting on sections has arrived, then contraction is necessary if the colony is to be worked for comb honey. But when a colony completely fills its hive, and has its combs well-filled with brood, I doubt if much is gained by contracting the brood nest. So long as the combs are kept full of brood, the surplus will go into the supers. If any of the combs of brood are taken away, they must be cared for by other bees somewhere else, so nothing is gained.

It is at the time of hiving a swarm that I have found contraction of the brood nest advisable. Years ago some of the "big guns" in apiculture were given to lamenting the swarming of bees, because, they said, with the swarm went all hopes of surplus. As the business was then conducted, the "big guns" were correct in many instances. The swarm would be hived in a ten-frame hive, and no supers put on until the hive was filled. If they *had* been put on they would not have been occupied until the lower hive was filled; and by the time this was accomplished it often happened that the white honey harvest had passed. If the old colony did not swarm again (usually it did), some return might be expected from that, unless the season was nearly over. In most of our Northern States the crop of white honey is gathered within six weeks, often within a month. If a colony is in a condition to begin work in the supers at the opening of the white honey harvest, and continues faithfully at work without swarming, as I have already said, no contraction is needed; but, suppose the harvest is half over, the bees are working nicely in the supers, there may be one case of sections almost ready to come off, another two-thirds finished, and a third in which the work has only nicely commenced, now the colony swarms, what shall be done? By hiving the swarm in a contracted brood chamber upon the old stand, transferring the supers to the newly hived swarm, and practicing the Heddon method of preventing after-swarming, work will be resumed and continued in the supers without interruption, and the surplus will be nearly as great as though no swarming had taken place.

When the brood nest is only one tier of frames, the only way by which it can be contracted is by taking out some of the outside combs, and filling the space thus left, by using "dummies." A "dummy" is simply a brood frame with thin boards tacked upon each side. It hangs in the hive and occupies space the same as a comb, only it is a "dummy" just as its name indicates. A frame wider than a brood frame may be used, and this will make the "dummy" thicker. Don't have the "dummy" touch the sides of the hive, then the bees cannot glue it fast. How thick a "dummy" should be depends upon how many combs are to be removed. When using the Langstroth frame I prefer to contract to five frames.

With the Heddon hive, in which the brood chamber is horizontally divisible, simply using only one section of the brood nest contracts the brood nest to about the proper capacity. This method of contraction is preferable to using dummies. Not only is there less labor and complication, but the flatness of the brood nest, and the absence of any dummies under the outer sections, make the bees more inclined to work in the sections.

When the brood nest is very much contracted, it has a tendency to cause a newly hived swarm to "swarm out" and leave the hive. When there is trouble from this source, the brood nest may be used nearly or quite full-size for two or three days, until the swarming fever has abated, and the bees have settled down to steady work. If newly hived swarms begin "swarming out," when I am using the new Heddon hive, I use a full-size brood nest for three days, and then shake the bees from the lower section of the hive, and use this section for the *upper* section of the next hive into which I put a swarm.

It has been urged against contraction that it results in small colonies at the end of the season. If it is carried to too great an extent, and too long continued, it certainly does. If a man wishes to turn bees into honey, so to speak, contraction of the brood nest will enable him to accomplish his object. If colonies are too weak in the fall as the result of severe contraction, they must be united; but the course pursued by nearly all who practice contraction, is to enlarge the brood nest again in time for the colony to build up sufficiently for a fall flow of honey, if there is one, or to become strong enough for winter. When bees are wintered in a repository of the proper temperature, I have never found that unusually populous colonies were any more desirable than smaller ones. This is one advantage of cellar-wintering, the population may be reduced to the minimum during the consumptive, non-productive part of the year.

The Use and Abuse of Comb Foundation.

THAT comb foundation has been a boon to bee-keepers, no one doubts; that money expended in its purchase is often returned many fold is equally true; but such is not always the case. All through the working season wax is being secreted to a greater or less extent. If not utilized it is lost. Of course, bees that fill themselves full of honey and hang in clustering festoons secrete wax to a *very much* greater extent than those engaged in bringing in honey. The bees of a swarm will nearly always, if not always, be found with large wax scales in the wax pockets. Having found that foundation is used at a profit in some places and at some times, the bee-keeping world seems to have decided, with almost no experiments, that bees ought never to be allowed to build comb naturally.

Years ago I practiced hiving swarms upon empty combs, upon foundation, and upon empty frames—empty except starters of foundation. The first swarm was hived upon comb, the second upon foundation, and the third upon starters only. This order was continued, the first year it was tried, until fifteen swarms were hived, when the use of empty combs was discontinued, as it was only too evident that they were used at a loss. I have reference here to what was used in the brood nest in hiving swarms when raising comb honey. The difficulty with drawn combs is just this: Before the queen will lay in old combs, the cells must be cleaned out and "varnished" or polished until they shine; and long ere this, especially if there is a good flow of honey, they will be badly needed, and will be used, for storage. In other words, combs are ready for honey

before they are ready for eggs, and the bees fill the combs at once
with honey, when, from some perversity of bee-nature, work, in
many instances, comes almost to a stand still. Having filled the
body of the hive, the bees seem disinclined to make a start in the
sections. Where bees *commence* storing their surplus, there they
seem inclined to continue to store it; and let the bees once get the
start of the queen by clogging the brood nest with honey, and that
colony becomes practically worthless for the production of comb
honey.

Bees Secreting Wax and Building Comb.

The advantage of full sheets of foundation over starters, or *vice
versa*, was not so apparent, and, until the close of the season, an
equal number of swarms were hived alternately upon full sheets of
foundation and upon starters. Enough was proved the first season
to show that, so far as surplus was concerned, nothing was gained
by using foundation in the brood nest, except for starters, when
hiving swarms. I have since continued to experiment, year after
year, by hiving swarms alternately upon foundation and upon starters only, in the brood nest, weighing both surplus and brood nests
at the end of the season, and the evidence has been in favor of empty
frames *every time*. Occasionally I have hived a swarm upon drawn

combs, but the loss has *always* been so great, that it seems like folly to repeat it.

When full sheets of foundation are used in the brood nest, and the brood nest is so contracted that some of the bees must enter the sections; and the sections are filled with drawn comb, or partly drawn comb, the honey must, from necessity, be stored in the supers until the foundation in the brood frames can be drawn out; and even then, having *commenced* work in the sections, the bees will not desert them. But there is only one queen furnishing eggs while hundreds of busy, eager workers are pulling away, with might and main, drawing the foundation out into comb; and the time eventually comes when there are thousands of empty cells in the brood nest. Now, Nature has no greater abhorrence of a vacuum than has a bee of an empty cell during a flow of honey; so, while the general orders are "up stairs with the honey," no cells in the brood nest are left empty very long. Especially is this true with a deep brood nest and yellow Italians.

If a swarm is hived upon starters only, the first step is, necessarily, the building of comb. If a super filled with drawn, or partly drawn comb (*not* foundation) is placed over the hive, the bees will begin storing honey in the combs in the super at the same time that comb building is begun below. A queen-excluder must be used to keep the queen out of the supers, then she will be ready with her eggs the moment a few cells are partly finished in the brood nest; and, if the latter has been properly contracted, she will easily keep pace with the comb building. The result is that nearly all of the honey goes into the supers, where it is stored in the most marketable shape, while the combs in the brood nest are filled almost entirely with brood. When bees are hived upon empty frames, a small brood nest is imperatively necessary, otherwise large quantities of honey will be stored therein; and when bees build comb to store honey, particularly if the yield is good, they usually build drone comb. So long as the queen keeps pace with the comb builders, worker comb is usually built, but if the brood nest is so large that the bees begin hatching from its center before the bees have filled it with comb, and the queen returns to re-fill the cells being vacated by the hatching bees, the comb builders are quite likely to change from worker to drone comb.

No fairer question could be asked than: What are the advantages of this system? In the first place, the cost of the foundation is saved; but, although this is a great saving, it comes about incidentally, as the non-use of foundation is only a means to an end, and that is the profitable securing of the greatest possible amount of honey in the most marketable shape; leaving the brood nest so free from honey that no extracting is needed when the time comes for feeding sugar for winter stores. Those who for any reason do not wish to use sugar, may still take advantage of this system by putting the unfinished sections back on the hives in time for the honey to be carried down and stored in the brood nest for winter. Or a case of brood combs may be put on over the sections as the harvest draws to a close, instead of putting on another case of sections. This will do away with nearly all unfinished sections, and the case of filled brood combs can be given the colony at the end of the season in place of *its* empty combs. By either plan, the number of *finished* sections is increased.

The greatest objection to this plan is that it cannot be depended upon to produce all perfect brood combs. I think I am safe in saying that I have had thousands of combs built under this management, and I think that at least eighty per cent. of them were as perfect as it would be possible to secure by the use of full sheets of foundation. A much larger percentage was perfect when I was using mostly the Langstroth frame, and contracted the brood nest to only five frames. This made the top of the brood nest, where the bees commence their combs, so small that the swarm completely covered it. All of the combs were thus commenced at the same time. As a rule, they were nearly as perfect as possible, at least so far as straightness was concerned. When I came to using the Heddon hive more extensively, I discovered that the greater surface at the top allowed room for the starting of more combs, that the outside combs would not always be started so soon as the center ones, and this sometimes resulted in the bulging of some of the combs.

Sometimes drone comb will be built in spite of contracted brood nests. Usually this is the result of old queens. But then, we can't always have young queens, hence I can only repeat that this method gives excellent results in the way of surplus, but cannot be depended upon to always furnish perfect brood combs. Some keep watch of the brood combs while they are being built, cutting out crooked, bulged or drone comb, and using it in the sections. I can not think favorably of such work. When I hive a swarm, I wish that to be the end of the matter. No opening of brood nests, and puttering with imperfect combs, during the hurly burly of swarming-time, would be desirable for me. But I do think favorably of contracting the brood nests when hiving swarms, then uniting colonies at the end of the season, culling out the imperfect combs and rendering them into wax. I think all such combs are built at a profit.

If securing straight, all-worker combs is not the greatest advantage arising from the use of foundation, it is certainly next to the greatest. The advantages of having each comb a counterpart of all the others, to be able to place any comb in any hive, in short, to have each interchangeable with all the others, and to be able to control the production of drones, to have them reared from such stock as we desire, and in such quantities, no more and no less, all these are advantages that cannot be ignored, even at the cost of filling our frames with foundation, and securing a little less surplus. We *must* have straight, worker combs. If they can be secured without foundation, well and good; if not, it must be used. By using weak colonies, or queen rearing nuclei, or by feeding bees in the fall, straight, all-worker combs may be secured at a profit.

Perhaps the greatest *immediate* profit arising from the use of foundation, is not so much in the saving of honey that would otherwise have been used in the elaboration of wax, as in the quickness with which it enables the bees to furnish storage for honey. When bees are storing honey slowly, the wax that they secrete without consuming honey expressly for that purpose, probably furnishes sufficient material, and there is probably abundant time, for the building of comb in which to store the honey. As the flow of honey increases, the handling of larger quantities of nectar increases the natural or *involuntary* wax secretion; but, as the yield of honey increases, a point is reached when honey must be consumed *expressly* that wax may be secreted. It is quite likely that, at this point, foundation may be used at a profit to aid the bees in furnishing storage. When the yield is so great that the bees cannot secrete wax and build comb with sufficient rapidity to store all of the honey that they might gather, then foundation is certainly used at a profit. Furthermore, I have seen the yield of honey so bountiful that even foundation did not answer the purpose; the bees did not draw it out fast enough to furnish storage for all of the honey that could have been brought in. At such times drawn combs are needed in the supers.

It will be seen that this question of foundation is one to which there may be profitably given much thought and experimentation. If the bee-keeper lives where the honey flow is light, but, perhaps, prolonged, he will find it more profitable to allow his bees to build their own combs. If he can't get perfect brood combs, he certainly can allow the bees to build their own combs for the surplus comb honey. And, by the way, no comb built from foundation can ever equal the delicate flakiness of that built naturally by the bees. If honey comes in "floods," as it sometimes does in some localities, the man who allows his bees to build their store combs unaided at such a time, loses dollars and dollars. If foundation is needed only for the sake of securing straight, worker combs, it need not necessarily be heavy. All foundation in brood frames, upon which swarms are hived, should be wired to prevent sagging and breaking down.

Copyrighted by H. E. Hill.

Orange Blossoms.

Increase, Its Management and Control.

THERE are two classes of bee-keepers who desire to prevent increase in the number of their colonies. The first, and by far the larger class, own large home-apiaries, and prefer surplus to increase. This class can allow *swarming*, if, by some simple manipulation, the number of colonies can be kept the same, and the bees induced to devote their energies to the storing of honey. The second class are the owners of out-apiaries; and while they may not be so particular about preventing increase, they do wish to prevent swarming. This accomplished, the out-apiaries can be left alone, except at stated intervals.

In reply to the question, "Why do bees swarm ?" it has been replied that "It is natural." "It is their method of increase." This may be true, in part, but it is not a satisfactory answer. I have never known a season to pass in which *all* of the colonies in my apiary swarmed or else didn't swarm. One year I had 75 colonies. They were worked for comb honey. Forty of them swarmed; thirty-five of them didn't. It would have been just as "natural," just as much "according to nature," for one colony to swarm as for another. In Gleanings for 1889 there was quite a lengthy discussion in regard to the causes that led to swarming. The chit of the discussion seemed to be that an undue proportion of young or nurse-bees to the brood to be nursed was the prime cause of swarming. If the brood-nest be well-filled with brood, then for lack of room the bees begin storing honey in the cells from which the bees are hatching, the result is that soon there is but little brood to care for, compared with the number of nurses, or young bees. This theory is strengthened by the fact that when bees are given an abundance of empty comb in which to store their honey, swarming very seldom ever occurs. In short, extracting the honey, or, to be more exact, giving plenty of empty comb, is the most successful, practical method of controlling increase. In large apiaries, especially out-apiaries that can be visited only at intervals, it is well-nigh impossible to keep every colony always supplied with empty combs, hence there will be occasional swarms. If there is to be some one present to hive what few swarms *do* issue, and prevention of increase is desired simply that the amount of surplus may be greater, and the surplus is preferred in the extracted form, then the man with these desires can have them gratified.

In the production of comb honey it is doubtful if there is a *prof itable* method of preventing swarming, although, of late, the practice of what is termed "shook-swarming" enables the bee-keeper to swarm a colony, in a manner very nearly approaching natural swarming, when he finds that preparations are being made for swarming. When he finds a colony building queen cells, he knows that within a few days, a week at the utmost, the colony will cast a swarm; and, instead of waiting, and allowing the colony to swarm when it has completed its first queen cell, he takes the matter into his own hands by shaking off most of the bees and the queen into a new hive, treating this shaken swarm in exactly the same manner as he would treat a swarm that had issued naturally. In other words, the bee-keeper simply forestalls what would have occurred naturally, in a few days, if the colony had been left undisturbed. The advantage is that the bee-keeper can thus bring about the swarming when he is present to attend to it, instead of having it happen when no one is present. This plan enables him to visit out-apiaries at stated intervals, giving each colony an examination, and "shaking" those that are making preparations for swarming. A colony that is not building queen cells is not likely to swarm inside of a week, and may be left undisturbed until the next weekly visit. Another minor advantage of shook-swarming is that it does away with the uniting and mixing up of two or more swarms that may issue at the same time in a large apiary, where natural swarming is allowed. Failures in shook-swarming result, as a rule, from doing the work too early in the season, before the colony has made preparations for swarming, and in not disturbing the bees sufficiently at the time, thus causing them to fill themselves with honey, as they do when swarming naturally. Before beginning the work, it is well to jar the hive, or pound upon it, until the bees are thoroughly frightened, and have filled themselves with honey. If it is desirable to have increase, the old hive can be given a new location and a laying queen, or a ripe queen cell. If no increase is desired, the old hive can be set by the side of the new one, with its entrance turned slightly to one side; then, at the next visit, it may be shifted to the other side of the new hive, when the flying bees will enter the new hive. A week later it may be placed back upon the other side, and, at the next visit, three weeks from the swarming, the few remaining bees may be shaken out of the old hive. The shifting of the old hive, from side to side of the new hive, *may* be omitted, the old hive being left standing by the side of the new one until the three weeks have elapsed, when all of the bees may be shaken in with the new colony. The advantage of the former plan is that some of the hatching bees are sooner thrown into the new hive, where their work will be to the greater advantage of the owner.

To avoid all danger from after-swarming it is desirable to shake the combs quite clear of bees when making a "shook-swarm," and this sometimes results in chilled or starved brood. There is a way, however, to avoid this difficulty. Set the new swarm a little to one side of the old stand. The flying bees return to the old stand and care for the brood At night the hive containing the brood, and the flying bees that have returned and entered it, is picked up and carried to a new stand, and the "shook-swarm" placed upon the old stand. Of course, the old bees that are carried to the new stand gradually come back to the old stand, and join the "shook-swarm," but it is one or two days before they all get back, and, in the meantime, young bees are hatching out, and, by the time the old bees have all returned, there are sufficient young bees hatched to protect and feed the brood.

When natural swarming is allowed to the extent of first swarms, it is an easy matter to prevent the issuing of after-swarms in a home-apiary where there can be daily attention. The plan is very similar to the one just mentioned for preventing increase when practicing shook-swarming. When the season for surplus honey closes with clover or basswood, it is better not to try to secure surplus from both the parent colony and the swarm. Hive the swarm upon the old stand, transferring the supers from the old to the new hive. If the brood chamber of the new hive is not too large, work will be at once resumed in the sections. Place the old hive by the side of the new one, with its entrance turned to one side. That is, have the rear ends of the hives nearly in contract, but their entrances perhaps two feet apart. Each day turn the entrance of the old hive a few inches towards that of the new hive. At the end of the sixth day, the two hives should stand side by side. Practically, the two hives are on one stand. True, the bees of each hive recognize and enter their own home, but, remove one hive, and all of the flying bees would enter the remaining hive. Usually the second swarm comes out on the eighth day after the issuing of the first. Now, if the apiarist will, on the seventh day, about noon, when most of the bees are a-field, carry the old hive to a new location, all of the bees that have flown from the old hive since the issuing of the swarm, that have marked the old location as their home, will return and join the newly hived swarm. This booms the colony where the sections are, and so reduces the old colony, just as the young queens are hatching, that any farther swarming is abandoned. The old colony just about builds up into a first-class colony for wintering. If there is a fall honey flow, such a colony may store some surplus then. This method of preventing after-swarming, called the Heddon method, is not *infallible*. If a colony swarms before the first queen cell is sealed, the first young queen may not hatch until the old colony has been upon the new stand long enough for a sufficient number of bees to hatch to form a swarm; but, as a rule, this plan is a success. If an after-swarm *does* come out, I open the hive, while the swarm is clustering, cut out all of the queen cells, return the swarm, and that is the end of the swarming. If the bee-keeper desires no increase, he can pursue the plan just given until it is time to remove the old hive to a new location, when it may be shifted to the opposite side of the new hive, with its entrance turned to one side, then gradually worked back to the side of the new hive, as has been already explained, then, at the end of the week, shifted back to the other side, where it may stand another week, when all of the bees may be shaken out, and the hive and combs removed. What little honey remains in the combs may be extracted, or, if some of them are well-filled with honey, they may be saved to give any colony that is lacking in stores at the approach of winter.

There seems to be no good plan of allowing bees to swarm, and then preventing increase by uniting, without having an extra set of combs built for each swarm that issues, and the same may be said when shook-swarming is practiced, but I believe such combs are produced at a profit.

There is still another plan of preventing increase, besides that of merging the old colony into the new; it is that of contracting the brood nest of the newly hived swarm to such an extent that the end of the season will find it too much reduced in numbers for successful wintering, when it may be united with the parent colony.

Quite a number of bee-keepers have succeeded to their entire satisfaction in preventing after-swarming, also in preventing increase, while only a very few have succeeded in preventing swarming when working for comb honey. Probably the only *certain* method that has been used to any extent, in this country, is that of removing the queens just at the opening of the swarming season, leaving the colonies queenless about three weeks. Of course, queen cells must be cut out, at least once, during this interval. Although a few good men practice this method, I never could bring myself to adopt it—there is too much labor.

The man who is raising comb honey as a *business* will find it to his advantage to allow each colony to swarm once, if it *will*, (and no more) then make the most out of the swarm. Whether the swarm and the old colony shall be again merged into one, depends upon the desirability of increase.

A YOUNG BEE-KEEPER.

The Hiving of Bees.

NATURAL swarming, with its uncertainties, anxieties, and vexatious losses, is destined to eventually become a thing of the past. Methods of controlling increase, preventing it altogethea, or else doing the work artificially, will reach such perfection that swarming will be eliminated. Many bee-keepers are already forestalling swarming by some artificial method of increase, notably that of shook-swarming. No professional bee-keeper worthy of the name, any longer allows natural swarming, un-controlled, in a large apiary. The days have past when we can afford to allow several swarms, issuing at the same time, to join forces and make merry in the top of some tall tree. Even if swarming is allowed, the queens are either clipped, or else controlled by means of queen-traps in front of the entrances of the hives. Two or more swarms issuing at the same time may unite, and give trouble by attempting to enter one hive when they return, but there will be no loss of bees, nor climbing of trees. The bees will *stay in the yard*, and can be brought under control.

When swarming is allowed, I believe that the majority of ad-vanced bee-keepers now hive their swarms by having the queens' wings clipped, and allowing the bees to return to their old location, which they will do when they find the queen is not with them. Of course the queen *attempts* to follow the bees, and is found in front of the hive by the bee-keeper, who cages her, and sets the old hive to one side, replacing it with a new hive prepared for the occupancy of the swarm. When the bees return, they enter into the new hive, supposing it to be their old home, thus hiving themselves. While they are entering the hive, the queen is allowed to run in with them —and the work is done.

There is another method of carrying out this principle; that of catching the queen in a trap in front of the hive. The lower part of the trap is covered with perforated zinc, the perforations being of such a size that the workers can pass, but not the queen. When a swarm issues, the queen attemps to follow, and, eventually, finds, and passes through, a cone-shaped opening in the upper part of the trap. Here she finds herself trapped in another apartment, as the chance that she will find the narrow mouth of the cone, and return, is as one in a thousand. The use of the trap saves clipping the queen's wing, also the looking for her when the swarm is out, together with the possibility of her being lost. The objections to the trap are its cost, a slight hindrance to the bees passing out and in, and its interference, somewhat, with the ventilation of the hive. A trap placed over the entrance of a hive containing a newly hived swarm will prevent loss if the swarm attempts to abscond.

If only one swarm would issue at a time, there would be no difficulty at all in managing swarms with clipped queens. When two or more swarms come out at the same time, and no water is thrown between them, they are almost certain to unite. After circling about for awhile, the bees return. If each bee would return to its old location, all would be well; but when the bees of one swarm begin to go back, a large share of the bees in the air follow them. A few bees from each swarm, even if several swarms have united, will return to their respective homes, but the majority will "follow my leader." It is impossible to give any set rules to be followed in such emergencies. If only two swarms have united, the bees may be allowed to enter the new hive until it is estimated that one-half the bees are in the hive, when it may be set out upon a new stand, and given one of the queens, then another hive set upon the old stand and the rest of the bees be allowed to enter. It should not be forgotten that, as a rule, other things being equal, a bee is worth as much in one hive as in another. Some bee-keepers, when several swarms come to one place, take supers from other hives, where the bees are working none the best, and place them upon the hive where the bees are entering. As soon as the supers are full of bees they are returned to the hives from whence they were taken. A colony made unusually strong by uniting swarms, will store more honey, but will be no stronger at the end of the season.

Another plan of managing, where several swarms come out at the same time, is not to allow the bees to return to their old locations, but put the caged queens in baskets, each queen in a separate basket, and hang the baskets on the branches of a tree where the bees show a disposition to congregate. The bees soon find and cluster about the queens in the baskets. As soon as a queen is found with sufficient bees to form a good swarm, remove the basket to a shady place, and cover with a cloth. Then remove the next basket that secures the proper quota, and so on to the end. Or the bees may be allowed to cluster about a single queen in a single basket, then the cluster divided up, and each division furnished a queen.

When natural swarming is allowed in a large apiary, and there is some one in constant attendance during the swarming season, I know of no more satisfactory method of managing than by the use of a swarm-catcher. With this arrangement there is no catching of queens, no climbing of trees, nor mixing of swarms—the control is perfect. The catcher is a light frame-work, about three and one-half feet long, sixteen inches square at the large, or outer, end, then tapered down to about three by sixteen inches at the small end. The outer end is closed with a removable door covered with wire cloth. The rest of the frame is covered with canvas or ducking. The small end is so made that it fits nicely to the entrance of a hive, and a portion of the cloth covering extends beyond the small

end, and forms a sort of flap that can be drawn over the mouth of the catcher, and fastened to keep the bees in after they have entered. In a large apiary there ought to be as many as half a dozen catchers scattered about the yard. When a swarm is seen issuing, a catcher is quickly adjusted to the entrance of the swarming-hive. In five minutes the whole swarm is in the catcher, when the catcher is closed and set in the shade, or carried to some cool place, like a cellar. The queen is usually among the last to leave the hive, so there is seldom a failure in catching her. If swarms come thick and fast, there is no objection to leaving the swarms several hours without hiving, provided they are not left in the sun. Although there is probably no necessity for it, they can be kept two days in a cellar. When the bees have been "cooled down" in this manner, and are then shaken down in front of the hive that is to be their home, they march in with scarcely a bee taking wing. Where some one can be in attendance, the swarm-catcher reduces the hiving business to an exact system.

While I do not approve of old fashioned swarming, with unclipped queens, in a large apiary, still, if a man *will* persist in following that plan, I will give him the best advice that I can; and, by the way, I can speak from experience, as I clung to that method until its disadvantages forced me to abandon it. When queens are allowed to accompany swarms, *water* is the great agent by which the bees can be controlled. Quite a number of pails filled with water should be kept standing in different parts of the apiary. There ought, also, to be three or four barrels of water standing about the apiary. Waiting one-quarter of a minute for water, sometimes means the loss of a swarm, For throwing the water, Whitman's fountain pump is the best. With this a stream of water can be thrown 30 or 40 feet. If two swarms issue at the same time, they can frequently be kept apart by the use of the pump. It is not necessary to throw a stream of water directly into the center of a swarm, but along one side of it, with a sort of sweeping movement of the arm, that makes the stream fall in a sort of shower. The bees dislike water and edge away from it. In this way they can be driven in any direction. Two or three pails of water thrown in this manner upon a swarm seems to disconcert the bees, and they begin looking for an alighting place. If the operator once has a swarm well in hand, there is plenty of water, and he knows how to use it, it is well nigh impossible for a swarm to get away.

When queens are allowed to accompany swarms, there ought to be no tall trees near the apiary, as the swarms *will* cluster where it is difficult, even dangerous, to get them. It should be possible to reach the tops of all trees with a long, light ladder. If the tops of the trees can all be reached from a step ladder, so much the better. Besides the pails of water, the fountain pump, and ladders, the bee-keeper will need three or four baskets. Clothes baskets are excellent. Upon one side should be sewed a cover of burlap. When the swarm has been shaken into the basket, the cover can be thrown over the top of the basket, and will keep the bees from flying out. Blocks of wood nailed to the corners of the cover hold it from being blown off or from dropping into the basket, should the bees cluster upon the cover. If set in a cool place, a swarm may be left in such a basket several hours. When the hive is in readiness to receive the swarm, the cover to the basket may be turned back, and the bees shaken down in front of the hive. A few of the bees soon find the entrance, and set up their "call" of a home is found, when the others follow them into the hive. If another swarm comes out, and attempts to join the one just entering its hive, a large sheet may be thrown over the hive.

Where several swarms come out at the same time and unite, the best that can be done is to divide them up as nearly equal as possible, into several swarms. When a queen is found she is to be caged. Any swarm that has no queen will soon show its queenlessness by its restlessness. The bees will begin running out of the hive and taking wing. One of the caged queens should then be given such a swarm, when, as by magic, the bees will change their behavior and go into the hive.

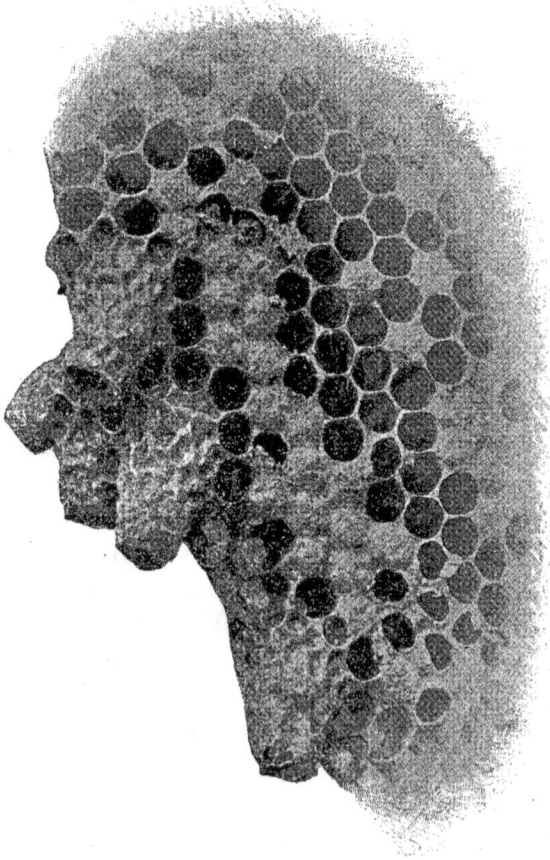

Cluster of Naturally Built Queen Cells From One of
Which a Queen Has Hatched.

Commercial Queen Rearing.

COMMERCIAL queen rearing is most decidedly a distinct branch of apiculture; as different from honey production as one profession is from another. *Time* is a most important factor in the rearing of queens; hence it is only in the South, where the seasons are long, that the business can be carried on at the greatest profit. It has been almost entirely abandoned in Northern States, as a man with a good location for honey can make more money producing honey than he can rearing queens. He might make more money per colony rearing queens than in honey production, but he can care for so many more colonies when they are worked for honey, that there is no comparison between the two in the short seasons of the North. A resident of the North occupying a location affording a light but continuous flow of honey throughout the season, might be justified in rearing queens instead of producing honey, but he could never rear the number of queens that he might rear in the South, simply because the seasons are not long enough.

The income from queen rearing is more of a *certainty* than that from honey production. If the blossoms yield sparingly, no surplus can be secured, but nearly, or quite, as many queens can be reared. Although a steady, moderate flow of honey is the most desirable for queen rearing, yet queens can be reared at a profit by feeding the bees.

In no branch of apiculture has there been such decided changes and improvements, in the last few years, as in that of queen rearing. Instead of scattering clusters of queen cells built upon some irregularity of the comb, artificially made cells are attached in a long row to a stick, supplied with tiny larvae, built out and cared for in a colony that has not even been deprived of its queen; and, best of all, the fertilization of a queen is now secured by using a mere handful of bees, where once were used at least two full-sized combs and a quart or two of bees.

For making the artificial cells there is needed a "dipping-stick," which is a round stick, 5-16 of an inch in diameter, with a peculiar taper at one end. The tapering part should be about 5-16 of an inch long, reduced rapidly for the first ⅛ of an inch, and then gradually reduced to the end. It should slip into a worker cell ⅛ of an inch before filling the mouth of the cell. These dipping sticks can be made with a lathe, from any kind of hard wood. Heart-cedar is best, as the water is slow to penetrate it, consequently, there is a little swelling. To dip the cells, beeswax must be kept just above the melting point by placing the dish containing it over a lighted lamp. Keep a little water in the dish, as this will be a guide to the temperature. No bubbling should be allowed. The stick, after being thoroughly soaked in water, is dipped rather less than ½ inch deep into the wax; four dips usually completing the cell and attaching it to the wooden bar upon which it is to be supported while in the hive. Dip three times, then loosen up the cup on the stick, then dip again, and immediately press the base of the cell upon the stick at the point where it is desired to have the cell remain. The cooling of the wax attaches the cell to the stick. If the stick or slat to which the cells are to be attached is soaked in melted wax until the frying or bubbling ceases, before attempting to attach the cells, they will adhere much better. If the queens are to be allowed to hatch while the cells are still attached to the stick, they being protected by some sort of a nursery, then there must be some kind of notches, or marks, on the stick to guide the operator in getting the cells attached at exactly the right places. An excellent method of managing this part of the business is to have a whole "battery" of dipping sticks thrust through holes bored at regular intervals in a wooden bar, when the whole row of cells can be dipped at only one operation. Five of these "batteries" can be kept in operation at one time; by the time the last one has been dipped, the first one has cooled sufficiently to be dipped again. After the fourth dip, while the wax is still hot and soft, the bases of the cells are pressed down upon the stick. As soon as the wax has cooled enough so that the cells will stick to the wooden slat, the slat is lowered into the wax until the wax covers it from end

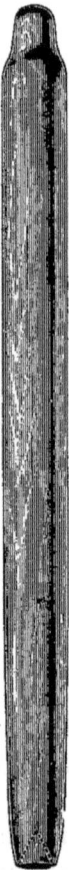

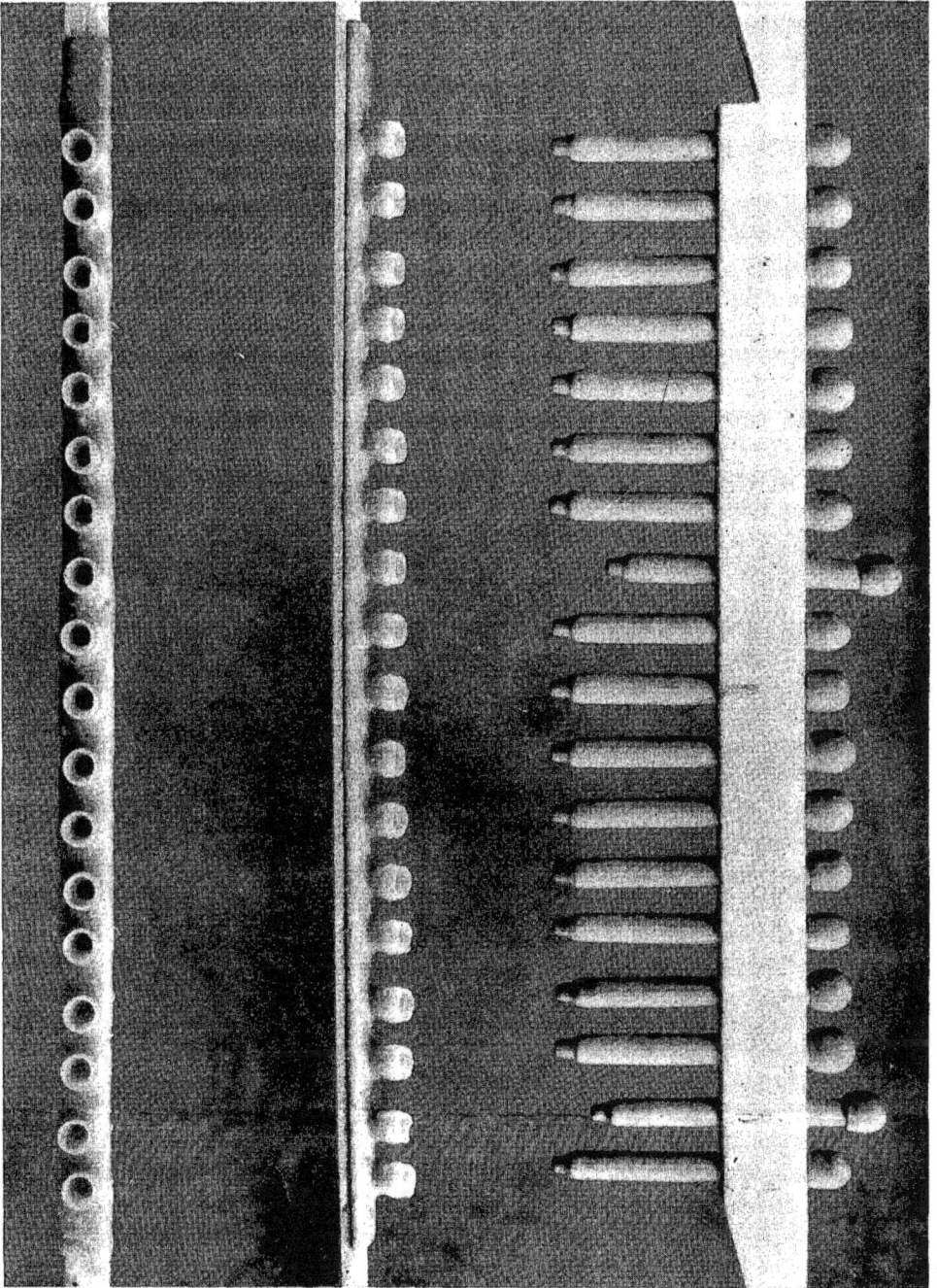

A Whole Battery of Dipping Sticks—and the Results.

to end, then dipped into water to harden the wax sufficiently so that the wax will not twist when the sticks are removed. Each stick is removed separately by turning it back and forth while it is being slightly withdrawn.

Right here let me say that much of the material in this chapter is taken from articles furnished the Bee-Keepers' Review by Mr. W. H. Pridgen, of Creek, North Carolina, and his plan of furnishing these cell-cups with young larvae is that of transferring the lining to the bottom of a cell, with the larva lying undisturbed upon the lining. Somebody has called it "taking up the baby, cradle and all." To make a success of this, the comb must be old enough so that the outside of the cocoon is black and glossy. By shaving down the cells with a keen edged knife, slightly heated, until the walls of the cells are only about ⅛ of an inch in depth, it is an easy matter to remove a cocoon with the accompanying larva. In fact, by bending the piece of comb back and forth the cocoons can often be forced to drop out of their own accord. To take up these tiny larvae, and transfer them to the cups, nothing is better than one of the dipping sticks already described. By making a little funnel shaped cavity in the opposite end from that used in dipping the cells, the larva and cocoon can be lifted by pressing this cavity down over them, much as a gun cap is pressed down over the tube. After placing the end of the stick in one of the cups, a slight pressure and a little twist, leaves the cocoon snugly ensconced in the bottom of the cell-cup.

There will be better success in having the cups accepted, and better results will be secured, if they first be given to bees deprived of both queen and unsealed brood from six to twelve hours previously. Nursing then begins the moment that cells are given. There are several methods of securing such conditions, but one excellent plan is to fill a hive with combs of brood, and set it over another colony, putting a queen excluder between the two stories. After the brood is all sealed in the upper story, it is placed upon a new stand, the queen cells cut out (if any there are), and one or two combs removed to make room for the frame in which is fastened the row, or rows, of prepared cells. After the bees have been left queenless for a few hours, they are ready to accept the cell-cups *instantly*. If allowed to do so, these bees would, of course, go on and complete the cells, but, after the work is nicely started, it has been found that they and the adhering bees may be placed in the upper story of an ordinary colony, when the bees will go on and complete them, *provided,* the queen is kept in the lower story by means of a queen-excluder. It is a singular fact that if one portion of the brood nest of a colony of bees is partitioned off with a queen excluder, the bees in that portion shut off from the queen will proceed to construct cells and rear queens. The only objection to placing the cell-cups there when first supplied with larvae is that the bees might be too slow in accepting these cells and feeding the larvae, and the result would be inferior queens. After the work is once started by queenless bees, as just explained, then these other bees will at once carry on and complete the work. After the just-started cells have been removed from the hive of queenless bees and gives to another colony, the hive of queenless bees may be set back upon the colony from which it was taken, the queen excluder left between it and the lower story, when it will be ready for starting a new batch of cells by simply setting it upon a new stand several hours before the prepared cups are given to it. In ten or twelve days all of the brood will have hatched in this set of combs, and another set must be started in time to be ready to take its place.

As the time approaches for the hatching of the queens, the cells must be removed from the bees, or protected in some manner, otherwise, the first queen that hatches will, with the assistance of the bees, tear down the other cells and destroy their occupants. Years ago, many queen breeders used what was called a lamp-nursery, that is, a hive, or box, with double walls of tin, the space between the walls being filled with water which was kept between 90 and 100 degrees by means of a lamp. The use of this device has been abandoned for the simpler and more reliable plan of leaving the cells with the bees, but protecting each cell by means of some mechanical device.

Mr. Arthur Stanley of Dixon, Illinois, attaches the cell-cups to the round, card-board gun-wads, one cup to each wad, then attaches the wads to a stick in something the same way as Mr. Pridgen attaches his cups to a stick, then, when the cells are nearly ready to hatch, he detaches the wads from the stick, and puts each cell into a small cylinder made of queen excluding metal. These filled cylinders are placed in rows, between two sticks (slipped in through holes made in the upper slat) and left in charge of the bees. The bees have access at all times to the cells, and to the queens when they hatch, but the size of the latter prevents their passing out through the openings.

Mr. Pridgen makes a nursery by taking a piece of board $\frac{5}{8}$ of an inch thick, two inches wide, and as long as a top-bar of a brood frame, cutting out a long notch, from one edge, a little more than an inch in depth, and nearly the whole length of the board, tacking wire cloth on each side, and dividing off the space between the wire cloth by means of tin divisions. These tin divisions are $\frac{5}{8}$ of

Cell-Cups and Finished Cells.

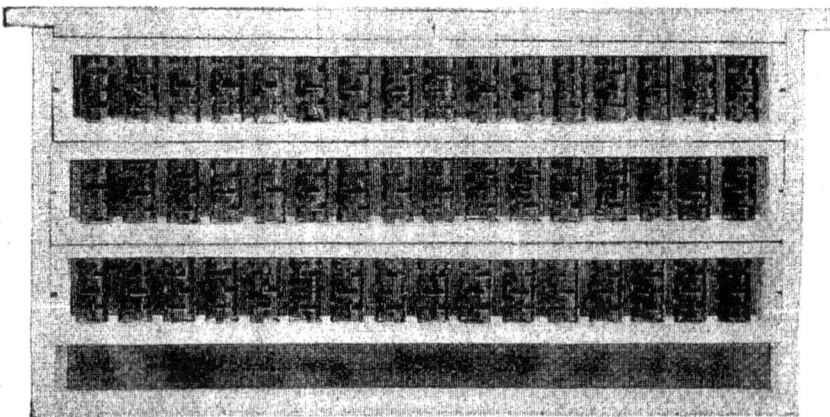

Cages of Queen-Excluding Metal.

STANLEY INCUBATOR AND BROODER.

an inch in width, a little more than an inch in height, and are let down slightly into saw kerfs cut in the upper surface of the wood. The divisions are kept in place by little points of tin that project from their upper corners through the meshes of the wire cloth and are then bent over or clasped against the wire cloth. Below each little cage thus formed is bored a small hole, through the wood, and in the hole is placed a plug having an opening hollowed out in its upper end and filled with soft candy. As the cells are all built exactly $\frac{5}{8}$ of an inch apart, in a straight row, attached to a stick, it will be readily seen how easy it is to lower all of these at once into the nursery, each cell occupying a cage $\frac{5}{8}$ of an inch square and a little more than an inch in depth. As the queens will all hatch within a few hours of one another, the cells can be left in the nursery until the queens have all hatched, or, even longer, if necessary. To remove a queen, pull out the plug below her cage, when she will crawl out and may be caught and put into a cage, or where ever desired.

Nothwithstanding the great improvements made in securing the building of queen cells, and in caring for virgin queens, they are no more wonderful than the late methods of securing the fertilization of queens—the most expensive part of commercial queen rearing. Once it was necessary to have at least two, full-sized combs and a quart or two of bees for each nucleus; now the fertilization of a queen is secured by the use of not more than 200 bees; one colony furnishing sufficient bees to secure the fertilization of 200 or 300 queens. This plan allows the introduction of queens from five to seven days old, and does away with all trouble from laying workers.

It is a little difficult to say to whom belongs the credit for this new method of caring for queens while being mated. Mr. C. B. Bankston, of Texas, published the first description that I remember having seen; but he had, at this time, a partner, a Mr. John W. Pharr, who says he helped in the development of the idea. Mr. W. H. Laws, of the same State, also helped in perfecting the plan and in bringing it to public notice. Mr. E. L. Pratt, of Swarthmore, Pennsylvania, has also done his share in making a success of mating queens from small nuclei. Like many other inventions, it seems to have been the work of several men. Mr. W. H. Laws published a description of the plan, in the Bee-Keepers' Review, and it is mostly from that article that the following is taken.

The nucleus boxes or hives play an important part in this system of mating queens, yet they are very simple in construction. Imagine two little trays, $\frac{3}{4}$ of an inch in depth, $11\frac{1}{4}$ inches long, and $4\frac{1}{4}$ inches wide, hinged at the bottom with leather strips, and made

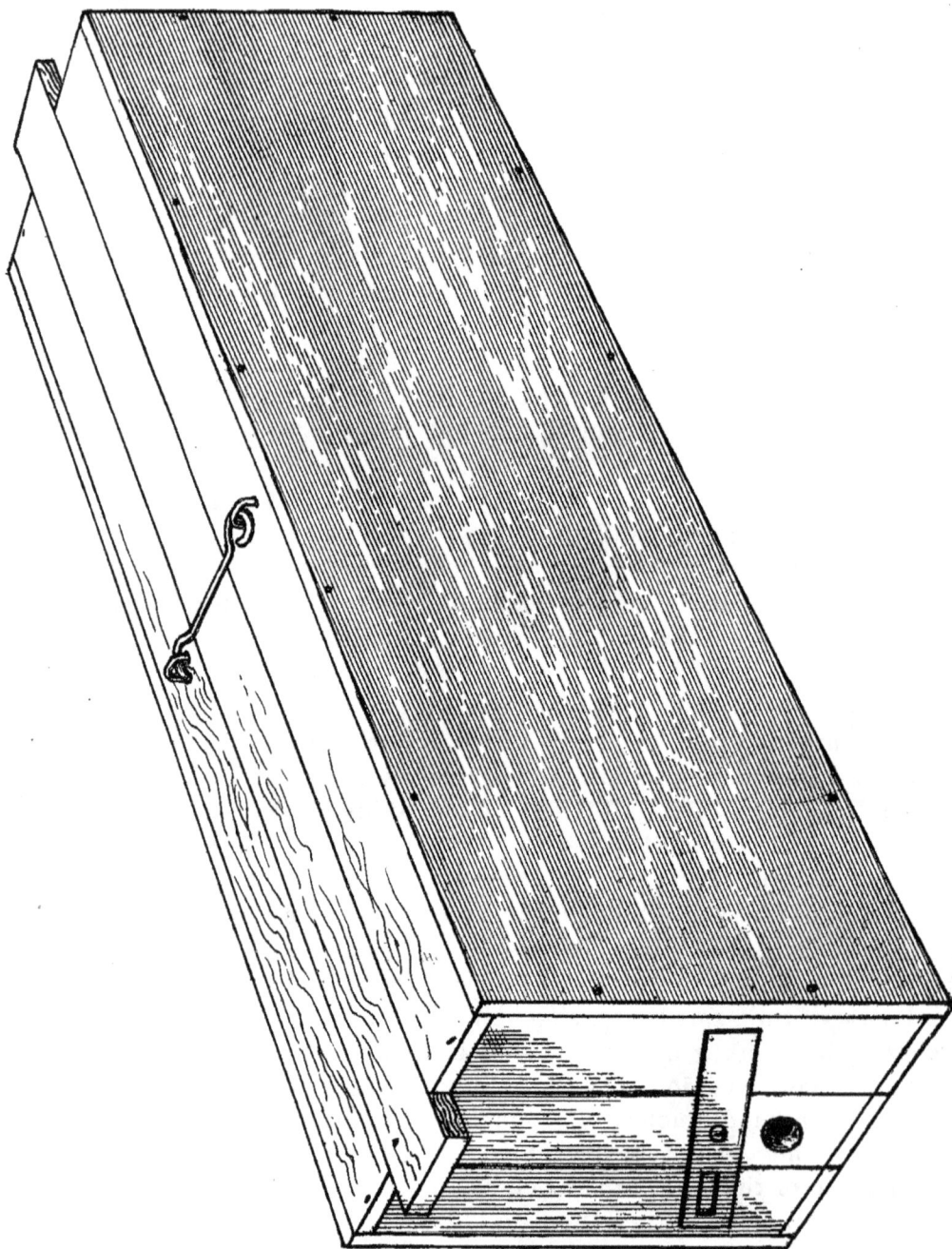

A "Baby" Nucleus.

so as to clamp a frame of honey (of the same dimensions) between them so tightly that it can not move. The outside surface of the frame becomes a part of the box, and all is held fast by a spring hook and staple at the top. An entrance for the bees is made in the end-bar of the frame by boring a hole with a 5-16 bit. A little button of sheet-zinc is fastened near the entrance. One end of the button is solid, closing the entrance, while the other end has a single, queen-excluding slot. When the button is turned crosswise, the entrance is left wide open.

To get these little combs of honey for the baby nuclei, frames are made to fit crosswise in an 8-frame, half-depth super, and old combs are transferred into them and given to ordinary colonies during a honey flow. Of course, sheets of foundation may be used, but old combs, well-attached, are preferable.

Equipped with as many of these boxes, already prepared, as we have virgin queens in our nurseries, we proceed to shake all of the bees from the combs of a populous, *queenless* colony (after first making the bees fill themselves with honey), putting the beeless combs into an empty hive, and setting it upon the old stand, to which enough bees will return to care for the brood. The old hive, containing the honey-laden, queenless bees, is now moved to some shady, convenient spot, where, with a small tin cup, we dip from the cluster a small wad of bees, say about the size of an unhulled walnut, containing about 100 to 150 bees, never more than 200, open one of the little boxes, and pour the bees right into the box, upon the comb of honey, close the box, snap the hook, and lay it aside. We keep on dipping and filling until all of the boxes are filled. The bees being loaded with honey, dip nicely; and, not being able to climb the smooth, inside of the cup, they handle about like so many beans.

Soon the bees in the boxes are all buzzing and roaring, and thus. lamenting their queenlessness and confinement, when we are ready to introduce the virgin queens, which is done by running them into the 5-16 inch, round entrances to the boxes. When the virgins are all run in, and the entrances closed, the boxes may lie in the shade until the evening of the next day, or even 48 hours, and no harm will result. The bees, being queenless aud confined, always accept the virgin queen, regardless of her age, or from whence she has come.

Within 24 hours, the bees in each baby nucleus have concluded that escape is impossible, and, resolving that "what can't be cured must be endured," they accept the situation, together with the queen, and quiet down. Later the nuclei may be carried out 300 or 400 yards, and the entrances opened as the nuclei are scattered under the brush, lodged in the forks of trees, or pitched into the weeds—anywhere, in any position, only be sure they are in the shade, where they remain a few days until the queens are laying.

These little, miniature swarms with virgin queens behave very much like newly hived swarms. Queenless when caged, and re-maiming 24 hours with a virgin queen, every bee seems to consider the box as home, and one or two bees are always on guard at the entrance.

Cell-Cups, Completed Cells, and Queen :

To gain time, the virgin queens are, as a rule, several days old when introduced to the nuclei, hence are ready to fly the next day after the nuclei are distributed. The third day after, the little zinc slots can be turned over the entrances so as to allow the workers to fly, but to retain the queens, thus preventing absconding. As soon as another batch of virgin queens is ready, these little boxes, when emptied of bees, are ready to be re-filled and used as before.

Another advantage of this method is the ease with which queens may be mated to select drones. It is the transportability of the nuclei that allows this. A man can carry from 15 to 25 on his arm, like a load of stove wood; or hundreds of them may be piled into a spring wagon, together with a colony of choice drones, and carried

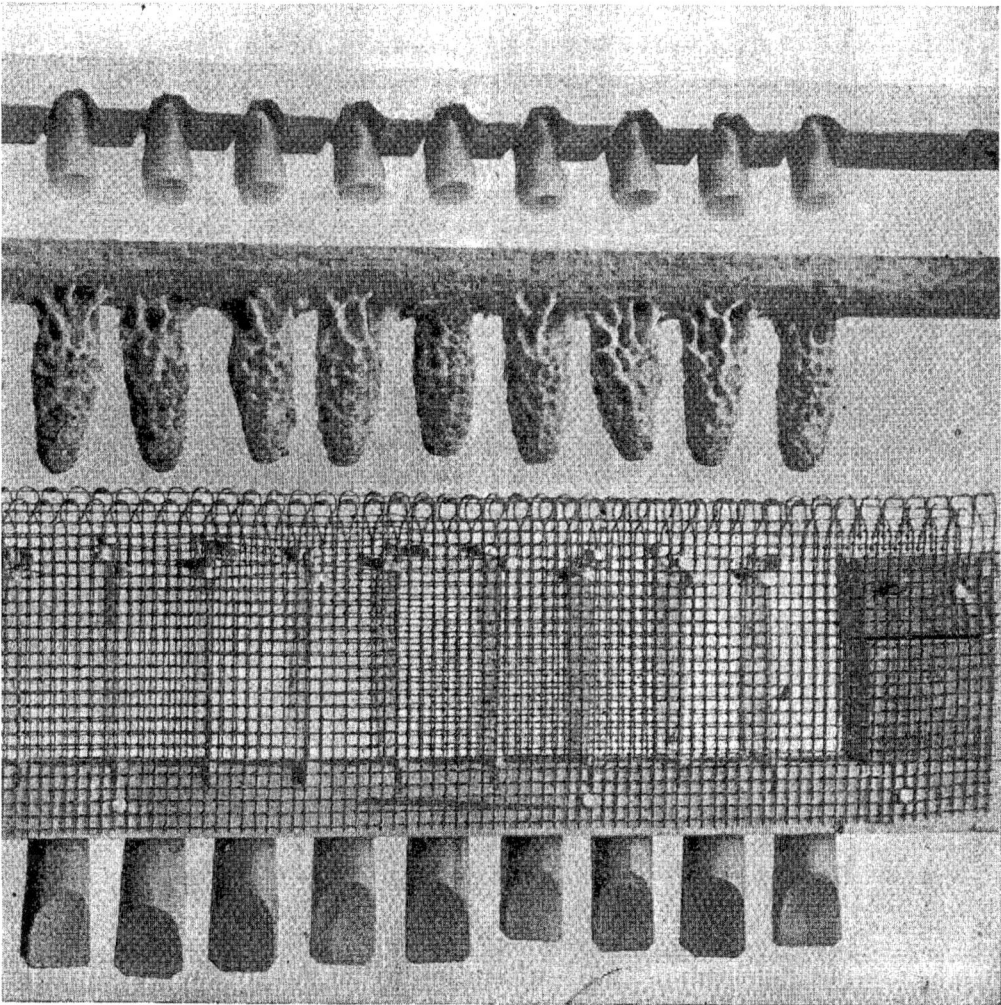

used by W. H. Pridgen, of Creek, N. C.

out three or four miles from any other bees. Possibly the next day every queen will mate, and that to the very drones of our choice, when all may be brought in, and, if so desired, queens and drones of *another race* may then be mated upon the same ground. Another thing, when the baby nuclei are carried away from all other bees, to the "mating grounds," no robbing occurs, even though there be a dearth of honey. Open air feeding could also be employed to advantage in such a location.

It will be seen that by this system there are never any permanent nuclei. A tiny cluster of bees is used to secure the mating of one queen, then the bees are shaken out (in front of the hive from which they were taken if the operator cares to take the trouble), and the boxes filled with a fresh lot of bees.

With all of the advantages of the "baby nucleus" system, there is one serious drawback, and that is of the nuclei being robbed at a time when no honey is coming in. During a honey flow, or if the nuclei are in an isolated location, this system is well-nigh perfection, but, under other conditions, the old-style, well-stocked nuclei have their advantages. It might also be added that while the old-style nuclei require more bees, they can also be managed with less labor after they are once established. I certainly consider the old system of sufficient importance to describe it.

In the first place, let me say that nothing is gained with any system by beginning queen rearing too early in the season. Wait until the weather is warm and settled and the colonies populous. In this latitude, May 10th is, as a rule, early enough to start queen cells. The first nuclei that are formed should be more populous than they may be used later in the season. Three combs are none too many to use at first; later these nuclei may be divided. About three or four days before the first batch of queens are to hatch, enough colonies should be made queenless to furnish bees for the nuclei, as queenless bees adhere much better to a new location. Many of the old bees will return, but, as most of the brood is sealed, enough bees (if they are queenless) will remain. When making up the nuclei, if the bees have been queenless, I would give each nucleus a cell nearly ready to hatch, or else a young queen, at the time of making the nuclei, as it seems to be something of an inducement for them to remain in the new location. As many bees return to the old stand, I leave some brood and honey in the hive, also put in some empty combs, and give the bees a laying queen. This colony soon builds up and prospers.

When a queen begins laying, she is allowed to fill the combs with eggs before shipment, then if a young queen is introduced soon after her removal, the nucleus receives another "sitting" of eggs in ten days more. By this management all nuclei are kept well supplied with brood.

When honey is coming in freely I have lost few queens by allowing them to run into the nuclei at the same time that the laying queens are removed. After a queen is two days old it is rather difficult to introduce her to these populous, well-established nuclei. Mr. Alley makes a success of it by smoking the bees with tobacco until they begin to show signs of stupefacation. The nucleus then has its entrance closed with a plantain leaf, as the bees are in no condition to defend their home. By the time the leaf wilts and releases the bees they are again able to defend themseves.

As a rule, a queen begins laying when ten days old, but hot weather and a good honey flow often shorten this period. I have frequently had them laying in eight days, and, in a very few instances, in only seven days. During a drouth, when no honey was gathered, I have known queens to be three weeks old before beginning to lay. At such times it certainly pays the queen breeder to feed the nuclei in which there are queens old enough to lay. When engaged in queen rearing I had some shallow boxes, each of which was just large enough to cover the top of the nucleus. These boxes were filled half full of candy, and when a nucleus was found during a dearth of honey, containing a queen old enough to lay, but *not* laying, one of these boxes of candy was inverted over the nucleus. Two days later, the queen would invariably be found laying.

With a large number of nuclei, it is impossible to remember the condition in which each was left at the last visit. A record of some kind must be kept, and I know of nothing better than the "Queen Registering Cards," sold by the A. I. Root Co., Medina, Ohio. They are made of weather-proof paper, and one is tacked upon the side of each nucleus. Upon each card are three dials. One contains the months; one the days of the month; and one the following words: "Eggs," "Brood," "Cell," "Hatched," "Laying," "Missing." Common pins are used as pointers. About ⅜ of an inch of the point is bent at right angles, and then driven into the center of the dial. When a cell is given, one pointer is turned to the month; another to the day of the month; and the third to the word "Cell." If the queen is found hatched at the next examination, the date is changed, and the pin turned to "Hatched." When found laying, and again when taken out and shipped, the pointers are turned accordingly. A glance at the register always shows the condition of the nucleus at the last examination. The turning of these pins takes only a moment, and is away ahead of using a pencil.

In the shipping of queens, success turns largely upon the food that is used. I have used no food superior to that obtained by mixing honey with pulverized sugar until it is of the consistency of a stiff dough. The novice is likely to make it too thin. It is well-nigh impossible to mix in too much sugar. The maker may think it quite hard and dry, but he will be surprised next day to find it quite soft and pliable. If too soft it will daub the bees and ooze out of the cage.

As a shipping cage I think the Benton stands at the head. It has three compartments all in a row and connected by openings. In one of the end compartments is placed the food; the other two are occupied by the bees, but the one next to the food is not ventilated, while the one in the end opposite to the food is freely ventilated. Very properly, this cage has been called the "climatic cage," as it enables the bees to accommodate themselves to the changes of temper- ature. If it is cool, they occupy the central apartment; if warm, they can remain in the ventilated part of the cage.

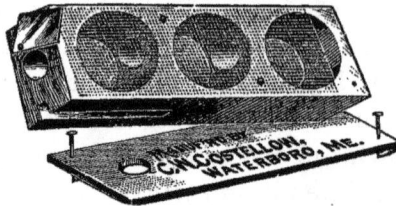

As to the number of bees to send in the cage with the queen, that depends upon the time of the year. In the spring and fall, more are needed; but I doubt if more than 30 bees are ever needed; in warm weather, one-third that number is sufficient.

The honey producer who wishes to rear a few queens for his own use will not find it worth while to follow many plans that are profitable in the hands of the professional queen breeder; he can divide up one or more colonies into nuclei, and supply them with cells taken from colonies that have swarmed, or he can remove a queen from a colony, and, when its brood has all been sealed, cut out the queen cells that have been started (as some of them may result in inferior queens), and give the bees a comb of eggs and just hatched larvae from some choice queen. The number of cells started will be increased if small holes are cut through the comb at that point where eggs are just hatching into larvae. When the cells are nearly ready to hatch they can be cut out, and one given to each nucleus.

Queen rearing does not call for any great outlay of physical strength, but consists rather of constant attention to a thousand and one little details.

Introducing Queens.

TO introduce a queen to a colony of bees, two things must be well-considered—the condition of the bees and the condition of the queen. The condition and behavior of the queen is very important. If she will only walk about upon the combs in a quiet and *queenly* manner, and go on with her egg laying, she is almost certain to be accepted if other conditions are favorable. Let her run and "squeal," utter that sharp "zeep, zeep, zeep," and the bees immediately start in pursuit. Soon the queen is in the center of a ball of tightly clinging bees, and the only course is to smoke the bees severely until they release the queen from their embrace, when she must be re-caged for another trial. Right here a caution: Don't hold the smoker too near the ball of bees, as *hot* smoke seems to infuriate the bees into stinging the queen. Hold the smoker far enough away so that the smoke will become cooled before reaching the bees. Dropping the ball into a cup of water has been recommended to induce the bees to release the queen; to the inexperienced, this may be the better plan, as it often happens that one of the bees will grasp the queen and endeavor to sting her, smoke or no smoke, and, in the attempt to rescue the queen, a novice is quite likely to injure her.

The Simmins method of introducing queens is an illustration of how great a part is played by the *attitude* of the queen towards the workers. He removes the reigning queen a few hours previous to liberating the new queen, and then, just at dusk, so late that the bees are through flying, and too late for the queen to take wing, the queen is released at the top of the hive and allowed to run down among the combs. And here comes in the important point: For half an hour before the queen is released, she is kept away from the bees and away from food, hence, when she comes in contact with the bees she is hungry and at once asks for food, instead of racing about the combs. The bees begin to feed and caress her, and all goes well. I believe Mr. Simmins claims that this method is almost infallible; and I remember that I once introduced ten queens by this plan without the loss of a queen. It was during a honey flow, however, and many plans that prove successful at such a time, may miscarry at times when no honey is coming in. The moral is to feed when trying to introduce queens during a dearth of honey.

To introduce a queen from one colony to another in the same apiary does not call for the skill needed when the queen has been absent several days from a colony, and is jaded by a long journey. I have frequently taken a queen from a colony, caged and sent it away, and then immediately taken a laying queen from a nucleus and placed her upon the spot upon the comb from whence I had taken the other queen, and had the satisfaction of seeing her immediately surrounded by a circle of admiring retainers. I believe there are times, particularly when honey is coming in freely, when a colony with a laying queen would accept *another* fresh laying queen, simply by having her placed upon the combs; and all would go well until the queens came in contact. Then there would be a conflict in which the chances of the new-comer would be equally as good as those of the old queen.

So far as the queen is concerned, it is important that she be brought before the bees in a natural manner; in such a place and in such a way as they would expect to meet her. When clipping queens, I have often replaced one in the hive by dropping her upon the tops of the frames, when the bees would immediately pounce upon her as an intruder. A puff of smoke would cause the bees to "let up," when the queen would walk majestically down between the combs, and there she was not molested, because there was where the bees expected to find a queen. When I wish to introduce a queen by allowing her to run in at the entrance, I first shake off the bees, from two combs, in front of the hive, and, as they are running into the hive, I allow the queen to run in with them. At such times there are no guards at the entrance, the bees that are crawling in will not attack the queen, and by the time the colony has recovered its tranquility, the queen is quietly parading the combs.

When a colony has been queenless long enough to build a batch of queen cells, I usually introduce a queen by simply taking a comb, with the adhering bees and queen, from a nucleus, and hanging it in the queenless colony. By means of smoke, or a feather, I drive all of the bees from one of the inside walls of the hive, and against this side of the hive I turn the side of the comb upon which is the queen. Then she is not immediately brought in contact with the excited, strange bees; but the bees intermingle, and, almost unconsciously, the whole colony accepts the queen. If any of the queenless bees stray near the queen, they find her surrounded by a cortege of her own bees. She is also attending to her duties, and is almost certain not to be molested.

When queens come from a distance, they are more difficult to introduce. They have not layed any eggs in several days, and are in a jaded condition. It would often be a saving in queens if such queens were first introduced to nuclei, and then, after they were nicely laying, introduce them to full colonies by uniting the nuclei with the full colonies, in the manner just described. It is much easier to introduce a queen to a nucleus than to a full colony. Take a frame of bees, brood and honey from a colony, place it in a hive with an empty comb by the side of the comb of bees, and set all on a new stand, when, in 24 hours, all of the old bees, the ones that always make trouble with a new queen, have returned to the old stand, and the young bees that are left are almost certain to accept a queen.

When a queen comes to hand in a jaded condition it would be a great advantage if she could at once be released upon the combs, but this very jaded condition is against her acceptance. To meet these conditions it is an excellent plan to cage her against the side of a comb. A cage for this purpose is made from a piece of wire cloth seven or eight inches square. First cut out, from each corner, a piece 1½ inches square, then ravel out several strands from each side, after which bend up the sides, at right angles, thus forming a sort of shallow box. The open side of this box is placed against the side of a comb where young bees are emerging, a few cells of honey also being included, the queen slipped under the edge of the cage, when the raveled out strands of wires are thrust into the comb, clear up to the cross-wires. Before doing this work it is well to free the comb of bees. The young bees that hatch will treat her kindly, and, in the meantime, she will begin laying, when, if the outside bees seem favorably disposed, the queen may be released by boring a hole

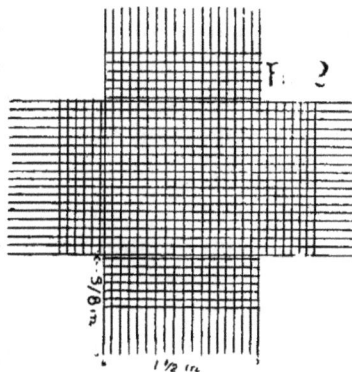

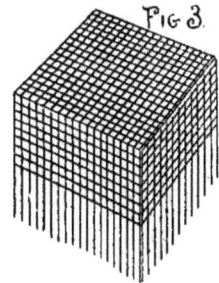

through the comb with the point of a pocket knife. The hole should be bored through from the side of the comb that is opposite to the cage, and, if the comb is simply broken up sufficiently, the bees will clean out the broken particles and thus allow the queen to pass through, which releases her at a time when the colony is in a normal condition, instead of under the excitement that accompanies the opening of the hive:

I just said that it would be well to release the queen if the bees were "favorably disposed" towards the queen. If they are "balling" the cage, clinging to its masses, like so many burdocks, their behavior indicates what the queen would have to endure were she within their reach. The operator must wait until they are in a different mood, until they are walking quietly about over the cage, as unconcernedly as upon the combs of honey—perhaps the bees may be offering food to the queen and caressing her with their antennae. This shows that the bees are favorably inclined towards the queen, and it is never safe to release a queen unless the bees show in this manner that they have practically accepted her.

Speaking of releasing the queen by boring a hole through the comb, reminds me that there is probably no better way of releasing a queen, let the cage be what it may, than that of stopping the entrance with a piece of broken up comb honey, or with some kind of soft candy, and allowing the bees to eat it out. The bees that first meet the queen are in good humor from the candy they have eaten; and, as has just been mentioned, the queen is released quietly at a time when the colony is undisturbed. After a queen has been released the hive should be left undisturbed three or four days, or a week, until the queen has commenced laying and become fully established as queen of the colony. When a queen has been released only a short time, she is easily frightened, when she is likely to run and "squeal," and the result is that the bees will at once "ball" her.

When a queen from a distance is to be introduced to a full colony, the *condition* of that colony is of the utmost importance. The most favorable condition is that it be hopelessly queenless. Let it build a batch of queen cells, and remove them after all of the brood has been sealed, and the bees are almost certain to accept a queen if given to them in a proper manner. When I was engaged in queen rearing, I don't know that I ever failed in trying to introduce a queen to a colony that had built a batch of cells. I would sooner release a queen after the bees had discovered the loss of their old queen, and before they had begun the construction of queen cells, than to release her after the cells were under way, *unless* I waited until the cells were sealed over and had been removed.

If the bees are shaken from their combs into a ventilated box, and kept confined, without a queen, several hours, Mr. Doolittle says that they will invariably accept a queen if given one in the box. In other words, they are hopelessly queenless, away from home, confined, and are ready to accept anything in the shape of a queen.

If the bees can, in some way, be placed in such a condition of mind (or body) as to let the queen alone until she has gathered the reins into her hands, so to speak, there is seldom any more trouble about her being accepted as their sovereign; and one excellent method of placing them in that condition is by the use of tobacco smoke. For several years I guaranteed the safe introduction of queens that I sent out, and the tobacco smoke method was the most successful of any that I ever asked my customers to try. The day before shipping the queen, I sent the following notice:

As soon as you receive this notice, remove the queen from the colony to which you expect to introduce the new queen. When she arrives, put her away in a safe place until after sundown. Just at dusk, light your smoker. When it is well to going, but in a pipeful of smoking tobacco, put on the cover, puff until you get an odor of tobacco, then puff two or three good puffs into the entrance of the hive. Wait two or three minutes, then send in another good puff or two, remove the cover, drive down the bees with a puff of smoke, open the cage and allow the queen to run down between the combs, following her with a puff or two of smoke, and then put on the cover. Half an hour later, light up the smoker again, putting in the tobacco as before, and blow two more good puffs in at the entrance. If no honey is coming in, feed the colony a pint of syrup each night from the inside of the hive, but don't disturb the brood nest for four or five days.

The tobacco smoke partly stupefies the bees, and by the time they have recovered, the queen is in full possession. By doing the work in the evening the bees are in condition to defend themselves by morning, should it be necessary.

There is, however, one method of introducing a queen that *never* fails, it is that of confining the queen in a hive with several combs of just hatching bees. Go over several hives, and find enough combs, from which the bees are just emerging, to fill a hive. Choose those combs having the least unsealed brood, as the most of this will perish. Shake off every bee, hang the combs in the hive, and close it up *bee-tight*. Allow the queen to run in at a small opening, closing it after her. This work should be done in the forepart of a warm day. In a few hours enough bees will have hatched to form quite a little cluster, with which the queen is *absolutely safe*. If the nights are cool, it might be well to carry the hive into the house for two or three nights. In five or six days the hive may be given a stand in the apiary, and the entrance opened sufficiently to allow the passage of a single bee. So much trouble as this is not advisable unless the queen is very valuable.

And now, in closing, a word of caution: When buying a queen from a distance, let out the bees and queen upon a window; catch the queen and put her into a clean cage; then kill all of the bees and throw them and the mailing cage into the stove. This is to guard against any possible chance of getting foul brood into the apiary from infected bees or honey. A queen has never been known to carry the contagion from one colony to the other—the only danger is that the food in the cage might have been made with honey infected with the germs of disease. Of course, the danger is very slight, even in this direction, but foul brood has been known to have been communicated in this manner, and there is no harm in exercising caution.

The Feeding of Bees.

BEES are fed to prevent them from starving when they lack stores in the winter, or in times of scarcity during the summer or fall, to stimulate the rearing of brood in the spring, or at any other time when it is desirable, to furnish them with winter stores when they are lacking in the fall, also to secure the completion of unfinished sections that may be left at the close of the honey harvest.

The feeding of bees for stimulating brood rearing in early spring is now looked upon by many as of doubtful value; especially is this true in the Northern States where weeks of warm weather are often followed by a "freeze-up." If the hives are well protected, and the bees well supplied with an abundance of sealed stores, natural brood rearing will proceed with sufficient rapidity, early in the spring, without any artificial stimulus; the only time that spring feeding is advisable is where there is a dearth of nectar, after the early spring flow and before the coming of the main harvest. A few bee-keepers have found it very profitable to feed enough at this time to keep brood rearing in progress, then, when the harvest comes on, the brood combs are full of brood and food, and the honey must go into the super instead of being stored in the empty cells of the brood nest. Not only this, but, as the result of uninterrupted brood rearing, great armies of workers are brought upon the stage of action at the proper time to help in the securing of the harvest. There come to my mind, now, two notable examples of men who have made a great success of this kind of feeding; one is H. R. Boardman of East Townsend, Ohio, and the other is Mr. E. W. Alexander, of Delanson, N. Y, Mr. Boardman uses a quart, Mason jar with a

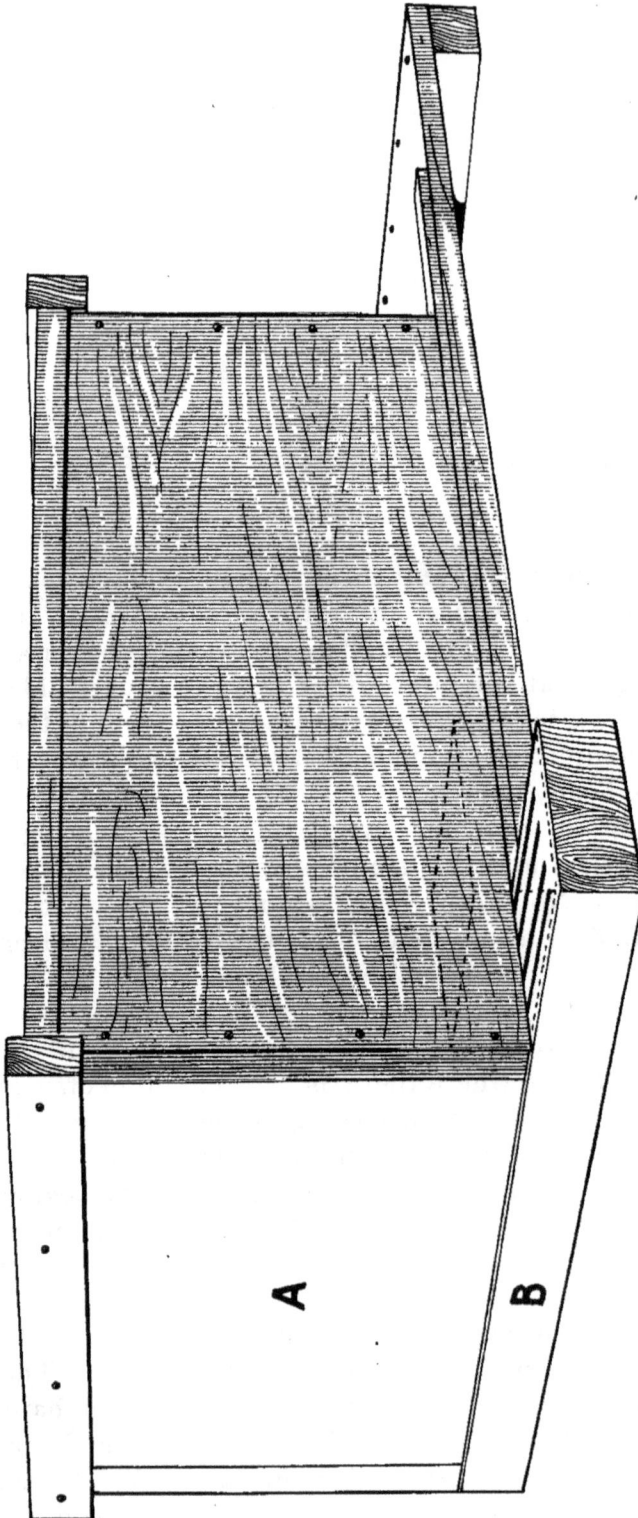

Bee Hive Arranged for Feeding with a Bottom Feeder.

A is the back end of the hive. B is the feeder in position. The dotted lines indicate the block used for covering that portion of the feeder where the feed is poured in.

perforated cover, the jar being inverted in a hole made in a shallow box that is placed in front of the entrance of the hive, the side of the box next the hive being open so that the bees can enter. The two side pieces of the box are made in such a way as to leave projections on their lower edges, on the ends next the hive, and these projections slip into the entrance, thus holding the feeder in place and making it more difficult for robbers to gain an entrance to the feeder.

Mr. Alexander, who, by the way, makes a success of keeping as many as 700 colonies in one apiary, makes a feeder out of a piece of 2 x 4 scantling about four inches longer than the width of the hive. With a cutter head, or a saw set wabbling, grooves are cut in its upper surface to within half an inch of the ends. This feeder is placed underneath the back of the hive, its upper surface on a level with the bottom board, the hive being shoved back on the bottom board sufficiently to cover the feeder. The feed is poured into the end of the feeder that projects out beyond the side of the hive, after which a block four inches square is laid over the projecting end to keep out robbers. When there are sufficient stores in the hive it is not necessary to feed so very much honey; a small quantity of food brought into the hive each day encouraging the bees to keep on breeding, using their sealed stores for this purpose.

Before feeding a whole apiary in this manner, year after year, I would suggest that the bee-keeper make an experiment: Feed one-half the colonies, keep an accurate account of the cost of feeding, and also an account of the net profit from each lot. Such an experiment, continued a few years, will answer the question as to whether such feeding is profitable in that particular locality.

Do the best we can with most methods of management, there will always be more or less unfinished sections left at the end of the season. What shall be done with these is really a serious question. If their number is not too great, those nearly completed may be sold in the local market, while the honey may be extracted from the remainder, and the bees allowed to clean them up by stacking them up in supers, out of doors, and giving only a small entrance to the pile of filled supers, when they may be used the next spring as "bait" sections to induce the bees to make an early start in the supers. If bees in large numbers are allowed to reach the sections while still wet with honey, they will, in their eagerness, tear down the cells and spoil the combs; for this reason, the entrance should allow only one or two bees to pass at a time.

When the local market is not sufficient to take the nearly completed sections, and there is a dearth of honey during the hot weather of August, it is possible to "feed back" extracted honey and secure the completion, at a profit, of all unfinished sections. I have fed back thousands of pounds of extracted honey for this purpose, and, for the benefit of those who wish to give the plan a trial, I will describe my methods.

As soon as I see that the white honey harvest is drawing to a close, which, with me, is about the middle of July, I remove all of the sections from the hives, look them over, take out the finished ones, and sort the remainder into three grades, viz., almost finished, half done, and just commenced. The cases containing the first two grades are then placed upon the hives, one case upon a hive, and allowed to remain until the bees have taken possession of them.

Then comes the task of selecting the colonies to do the work; and, by the way, this is the most important point of all. First, the colonies must be strong; next, they must possess young queens, preferably those of the current year, although this is not imperative; and, last, but not least, simon pure blacks are given the first choice. Hybrids are the next best, while, as a rule, Italians do very poor work in this line. Keeping these points in view, I select one-half as many colonies as I have cases of unfinished sections upon the hives, and to these colonies I transfer the cases—sections, bees and all—putting two cases upon a hive. I have never experienced the least trouble, in any respect, from thus mixing up the bees, while populous colonies are secured thereby.

If the brood nests are not already contracted, I contract them. The greater the contraction, the more satisfactory will be the results, so far as work in the sections is concerned, but, if carried too far, it will materially weaken the colonies by curtailing the production of brood. I have sometimes contracted the brood nest to only three Langstroth combs, and these three combs, when I was through feeding, were three solid sheets of brood; but, all things considered, I prefer to contract the brood nest to about the capacity of five Langstroth combs. There is also another point that must not be neglected, and that is that the brood combs must not be old and black, otherwise, the combs in the sections will become travel-stained unless removed very promptly upon their completion. The newer the combs in the brood nest, the better.

When honey is brought in from the fields it is carried up into the sections; that is, the supply, as regards the sections, comes from below; in feeding back, the feeder is usually placed above the supers, in which case the supply comes from above. In both instances, the sections in which the work is the least advanced should be placed nearest the source of supply. Thus it will be seen that, in feeding back, the sections that are nearly finished are placed next to the brood nest, and above these the grade that is about one-half completed.

The feeder that I used is the Heddon, which is exactly the size of the top of the hive, and is placed above the sections. His new feeder is unexcelled for this purpose, as the bees take down the feed from both sides. This might not seem important, but it is, and for this reason; when the feed is carried down upon one side only, the sections upon this side are completed *first*. When the feed is carried down from both sides, the sections are finished up very evenly all over the case. In this feeder, the reservoir is in the center, and just over it the cover slides back in grooves. There is no contact with the bees, no smoke is needed, no propolis is disturbed, and the cover fits so snugly that no odor of honey escapes to attract robbers.

The Heddon Feeder.

The bees seem to be able to handle the honey to better advantage when it is thinned somewhat, say, one quart of water to ten pounds of honey. I heat ten quarts of water over an oil stove until it boils, then mix it with 100 pounds of honey, stir it up well, when it is ready for use. The first feeding should be done at dusk, as it puts the bees in an excited state, and trouble from attempts at robbing might result. After the bees have become accustomed to finding honey in the feeder, feeding produces little, or no, excitement; still, at dusk is the best time to feed, as the annoyance of having robber bees follow from hive to hive, and dive into the feeder reservoir when it is opened, is thus avoided. The feed is given as fast as the bees take it.

Close watch is kept of the sections in the lower cases, and whenever a case is found in which all or nearly all of the sections are completed, off it comes; the case above it is placed next the brood nest, and above this case is placed a case of sections brought from the honey house, one containing sections of the third grade; that is, those in which the bees have made the least progress. I continue to bring in the the cases of finished sections as they are completed, replacing them with the unfinished ones from the honey house. When the stock of the latter is exhausted, I am ready to begin to reduce the number of colonies upon which I am feeding back, and this is done as fast as the sections are completed.

During all this time, since the feeding commenced, I have been watching each colony, and jotting down, upon the cover of the feeder, its characteristics; and, in reducing the number of colonies, those are rejected that have done the least satisfactory work. I continue to keep two cases upon each hive, and, as the colonies work with greatly varying rapidity, there is no difficulty, by changing about the cases, to keep next the brood nest those sections that are the nearest completion. In gathering the sections together upon fewer hives, I always take bees and all, thus I am continually strengthening the colonies upon which I am feeding back.

It is useless to expect the bees to finish up all of the sections upon a hive. Even though the feeding is continued, the sections will not be completed in a satisfactory manner. So long as the feeding is continued the bees seem to reason something like this: "We must make the cells as deep as possible, and delay the capping until the last moment, in order to make room for all the honey that we can; and, if there are not cells enough, we must build more, even if it be in the little, cramped up places between the tiers of sections." After the combs are drawn out to full length, filled with honey, and nearly sealed, I have secured better results by giving the bees no feed for three or four days, then giving them a light feed, and omitting the feeding for several days. The bees then behave as though they considered the harvest over and ended. They seal up most of the cells, and from those that they do not seal they remove the honey. But there is a much better way of managing the business. When the sections are all nearly finished, I put them upon as few hives as possible, and still not have more than two cases upon one hive, and then upon each hive, above the two cases of nearly completed sections, I place a case of sections filled with comb foundation. The bees proceed at once to draw out the foundation and fill it with honey, and this additional storage room appears to bring about a feeling that there is no further necessity for holding cells open below, and they are sealed, forthwith.

When the two lower cases are completed, the upper case (the one that was furnished with foundation) will, perhaps, be found to contain sections one-half completed, and these upper cases may be gathered together, bees and all, and placed, two upon each hive, over those colonies that have shown the greatest aptitude for this kind of work, and the feeding continued until the sections are almost completed, when it will again be necessary to place a case of sections containing foundation upon each hive. I have continued this operation until all the sections were finally upon one hive, and had all of the sections completed except those in the case last added on top.

After bees have been fed awhile, they secrete large quantities of wax. The little flakes of it can be seen between the scales of the abdomen, and, unless allowed to build comb, the bees will plaster with wax the woodwork of the sections, the inside of the feeders, cases, etc. The moral is to allow them to build comb. Have a row or two of sections in the upper case filled with starters only; thus there is secured, in the shape of comb, what would otherwise be wasted.

Although we cannot control the temperature, it may be well to know that the hotter the weather the more rapid and satisfactory will be the work of the bees when we are feeding back.

If there is any time when separators are needed, it is in feeding back. If the combs, both finished and unfinished, could be left undisturbed upon the hive, and the bees fed until all the combs were finished, feeding back would be no reason why separators should be used; but when the unfinished combs are put back in the cases, a great deal of judgment and patience are needed, unless separators are used. Bees usually leave a space of about ⅜ of an inch between combs, and, in putting back unfinished sections, where separators are not used, this fact must be kept in mind. When the space is less than this, no harm is done unless it is so small that a bee can not pass through, when the bees will connect the two surfaces by little bridges of wax, and when the sections are taken apart, these little, connecting bridges will pull pieces out from one comb or the other. When the space is much greater than ⅜, and the comb upon each side is sealed, the bees, especially if crowded, will construct comb upon the sealed surface of the other comb, which gives it a very botchy appearance. If the comb at one side of the space is sealed, and the other not, the sealed comb will be undisturbed, while the unsealed cells upon the other side will be lengthened out until the space between the two combs is reduced to about ⅜. If, in this instance, the sealed comb is smooth and even, and in the right place as regards the section as a whole, all will be well; but, if it be concave or convex, the unfinished comb facing it will be drawn out in conformity with the surface of the finished comb. If two unfinished surfaces, in the same stage of completion, are brought facing each other near the center of the super, they will be drawn out and sealed straight and true and alike; if they are near the outside, the chances are that the comb nearest the center of the super will grow faster than the one farther out, and a bulge will be the result. Combs near the center of the super are drawn out quicker and finished sooner than those at the outside and corners; hence I place at the outside those sections that are the nearest completion; and especially do I take pains to have sealed surfaces come next to the sides of the super, while combs that are the farthest from completion are placed in the center. By this management, all of the combs are finished at about the same time. Unless some of the combs begin to show signs of travel-stain, it is better to leave on the super until all, or nearly all, of the sections are completed, for, as the combs near completion, this matter of adjustment becomes more difficult. When separators are used, all of these troubles vanish.

When foundation is used, and comb honey produced, "right from the stump," so to speak, by the feeding of extracted honey, we have none of this patching, bulging difficulty to contend with, as all of the combs grow alike; and some of the finest, straightest, plumpest and most handsome comb honey can thus be produced that the eye ever beheld; but I have never found it profitable, except by placing a few cases on top, near the close of finishing up a lot of unfinished sections, to give the bees room, and thus induce them to seal up nearly finished combs, as has been already explained.

I know of only two objections to the feeding back of extracted honey. One is that "fed honey" has a slightly different taste from that stored directly in the combs from the flowers. There seems to be a sort of "off" taste, or lack of flavor. This lack of fine flavor is not very pronounced, but it can be noticed by those who are experienced in the matter. It is possible that this taste comes from the thinning of the honey and the handling of it about in different vessels, as well as the continued use of a wooden feeder. The other objection to "fed honey" is that it will candy much quicker than other honey. When the sections are nearly completed, and feeding is done simply to have them completed and sealed over, the proportion of "fed honey" is so small that these objections are not very serious. "Fed honey" ought to be sold early and in a market where it will be consumed before it candies.

Taking one year with another, I have secured about two pounds of comb honey from the feeding of three pounds of extracted. With the right kind of weather and colonies, I have done much better—secured four pounds from the feeding of five.

The advantages of feeding back honey can be stated in a few words: Comb honey is more salable, and at a higher price, than extracted, and if the latter can be changed into the former at no great expense, there are quicker sales and greater profits. The greatest advantage, however, is in securing the completion of nearly finished sections.

I think that the feeding back of extracted honey is on the wane, as bee-keepers are learning how to greatly lessen the number of unfinished sections at the end of the season; and my object in describing the practice is not to recommend it for general use, but to furnish the necessary instructions should circumstances arise making it desirable to follow them.

When bees require feeding in the fall, almost any kind of feeder will answer the purpose. If nothing better is at hand, a tin pan, or any open dish, may be set in an upper story, and a piece of burlap laid in the feed as a float for the bees to stand upon. A good-sized feeder, one that will hold from 15 to 20 pounds, like the Heddon, for instance, greatly facilitates the work, however. Bees can be fed with the Heddon feeder when it is so late and cool that no other feeder would answer. Fill the feeder with hot syrup, as hot as it can be and not burn the bees, then set the hive *over* the feeder, when the heat from the syrup will warm and rouse up the clustered bees, and they will come down and carry up the feed in short order.

The idea seems to prevail that all winter stores must be sealed. This is an error; and probably arose from the fact that late-gathered stores are often of poor quality—not because they may be left unsealed, but from the quality itself. A good, thick syrup made from granulated sugar is an ideal winter food whether it is sealed over or not; in fact, bees in a warm cellar may be successfully wintered on sugar syrup supplied to them daily by means of a feeder. A large, flat cake of candy laid over the cluster, and covered with enameled cloth, with packing of some kind over that, is a handier method of winter-feeding; but, aside from that, is not superior to the use of syrup, when the bees are in a warm cellar, but it would be out of doors.

In closing, let me caution the bee-keeper to beware whence comes the honey that he feeds. Let him be *sure* that it contains no germs of foul brood. To buy honey in the open market and feed it to the bees would be a most risky proceeding. For stimulative feeding, and for winter stores, better buy sugar. It is cheaper, safer and better—especially so for winter stores.

The Production of Comb Honey.

HAVING now considered some of the most important points in modern bee culture, such as locality, hives, supers, sections, increase, feeding, varieties of bees, use of comb foundation, etc., let us begin at the opening of the season, and go briefly over the ground, showing the relation of these different features to one another, as they are employed in the production of comb honey.

We will suppose that the early spring has passed; that the bees have received sufficient protection; been supplied with ample stores; and that the hives are now teeming with life as we stand upon the threshold of the main honey flow. And right here let me say that unless the colonies *are* strong and populous, simply overflowing with bees, it is folly to expect a paying crop of comb honey. If there is any time when weak colonies may be united to advantage, it is at the opening of the main harvest, when comb honey is to be the product. Better gather together, into one hive, three-fourths, or even all, of the bees and brood from two, three, or even four, hives, and thus have one rousing colony, than to attempt to secure a crop of comb honey with weaklings. A comb or two of brood and bees, and a queen, left in a hive at the beginning of the harvest, will build up into a good colony by fall, and, possibly, store some honey that may be extracted. No matter how it is accomplished, one thing is *imperative*, and that is that the brood nest be crowded with bees and brood at the opening of the honey harvest.

This condition tends greatly to make the bees begin promptly to store honey in the supers. And this is important, as, otherwise, the bees are inclined to crowd the brood nest with honey as the bees hatch out, also to "loaf," and develop the "swarming fever." If bees can be induced to begin working in the sections at the opening of the main honey flow, it relieves the "pressure," so to speak, upon the brood nest, which results in more brood, while the turning of the energies of the colony into the storing of honey, does much to keep down the swarming fever. The greatest attraction that can be placed in the supers is that of drawn comb. Unfinished sections saved over from the previous season are excellent for this purpose. As has been explained in a previous chapter, the honey must be extracted, and the bees allowed to clean up the combs, when the latter must be packed away in supers where no dust nor mice can get at them. I have given supers full of these partly drawn combs to colonies, (one super to a colony) and had these combs filled and capped, and ready to come off, just as other colonies supplied with sections containing foundation only, were only making their first start in the supers. In this case, a super of partly drawn combs was worth as much as a case of finished honey. There is, however, a still better method of managing this part of the business. It is that of putting on an extracting super first, and when this is filled and removed, the bees are always ready to go to work in the sections *immediately*. For this purpose, shallow supers are preferable; those containing frames half the depth of the regular brood frame being the size that is usually employed for this purpose. The greatest objection to the use of the full-sized combs is that it requires so much honey to fill a super of them, that it would materially reduce the crop of comb honey. The use of a shallow extracting super removes this objection. Again, the beginning of the white honey flow is sometimes mixed with an earlier, darker flow, and the taking of the first of the white flow in the extracted form, insures the perfect whiteness of all honey stored in the sections. Still further, these half-depth, extracting supers can be used to fully as great advantage at the end of the harvest as at the beginning—perhaps to greater advantage. As the time approaches for the close of the harvest, instead of giving more sections, simply set on top of the sections one of these half-depth, extracting supers. If more honey comes in than is needed to fill and complete the sections already on the hive, it will "overflow," so to speak, into the extracting super that is on top; thus the honey that would otherwise go to the making of a lot of unfinished sections, is secured in the extracted form. Getting extracted honey at the opening and closing of the season, as just explained, certainly has some very decided advantages. It leads the bees to begin work promptly in the supers at the opening of the season, keeps all "mixed" honey out of the sections, and practically does away with unfinished sections at the end of the season.

When the first case of sections placed on the hive at the beginning of the harvest is partly finished, it is raised, and another case placed between that and the hive. At what stage of completion the sections should be when a second case is added depends upon how crowded the bees are and the rate at which honey is coming in. I usually add another super when the sections in the one next the hive are from one-half to two-thirds completed. I have not found it profitable to tier up sections more than three supers in height. As a rule, the upper super is ready for removal before it is necessary to add a fourth. If it is not, and honey is coming in rapidly, I would transfer it, bees and all, to some other colony having a less number of cases, rather than tier up four cases high. With any system in which the sections are finished in close proximity to the brood nest, their removal is necessary soon after completion, to prevent their being soiled or "travel-stained," by the bees passing over them directly from the brood nest; but, with the tiering-up system, the finished combs are so far from the brood nest that they remain unsullied until a whole case can be removed at once. During a regular "honey-shower," such as we have *sometimes*, when the nectar all but drips from the fragrant, golden blossoms of the linden, I have seen a colony draw out the foundation in 28 sections, and fill them full of honey (and here is where I believe foundation is *very* valuable) in less than three days, yet scarcely a cell would be *sealed*. To give the bees another super next the hive is the work of only a moment. At such times it may be advisable to remove the upper case, after they have been tiered up three high, even if there are one or two unfinished sections in each corner; and, when crating, have an empty super at hand in which to put the unfinished sections, and when it is full place it on a hive.

When a super is ready to come off, there is no easier, less troublesome, method of freeing it from bees than by the use of a Porter bee-escape, which consists of a tin frame-work or box inside of which are two delicate brass springs so nicely adjusted that a bee can easily squeeze out between their points, but cannot return. Openings in the upper and lower sides of the box allow the bees to pass through. The escape is fastened into an opening cut in the center of a thin board the size of the top of the hive, a ⅜ rim around its edge holding the super bee-space above the board. To use the escape, simply raise the upper super, lay the escape-board upon the top of the next lower super, replace the removed super upon the top of the escape-board, and the work is done, so far as the bee-keeper is concerned. The bees, finding themselves shut off from the rest of the hive, become excited and make frantic efforts to

Sprig of Basswood in Bloom.

escape. Finding one opening by means of which they can reach "home," they crowd through as fast as possible, when, in a few hours, the super is free of bees. If escapes are put on at evening, the supers above them will be free of bees in the morning.

If there is not time to use escapes, or, if for some reason, it is not desirable to use them, the supers can be freed of bees by other methods. My practice has been as follows: Have the smoker in good trim, take off the cover, and drive a perfect deluge of smoke down among the bees. This starts them out of the combs at a lively rate, and, before they have time to come back, I have the super off the hive. The super is then tremulously shaken in front of the hive until most of the remaining bees are dislodged, when it is taken to the honey house and set on end. In a short time the few straggling bees leave the super and escape by way of the window, which should

have wire cloth over it on the outside, letting it extend several inches above the window, and terminate in a small cone-like opening from which the bees can easily find their way out, but not be very likely to find their way back. If the shaking process is found too laborious, and robbers are not troublesome (and they will not be until the close of the season), the super may be leaned against the side of the hive, near the entrance, when the bees will desert the super for the hive. When robbers are troublesome, the stragglers may be driven out with smoke, and brushed off in front of the hive.

By shading the hives, allowing generous entrances, also abundance of room in the supers, swarming is greatly delayed, and often avoided entirely with many colonies. I have known seasons when, with this management, not more than one-half of my colonies swarmed, and I have frequently had seasons when not more than two-thirds of them swarmed. When a swarm does issue, I hive it in a contracted brood nest, with starters only in the brood frames, on the old stand, put on a queen excluding honey board and transfer the supers from the old to the new hive. In 20 minutes, at the outside, the bees are back at work in the sections that they recently deserted in such a hurry. The old colony is placed by the side of the new one for a week, when it is moved to a new stand, thus throwing all of its flying bees into the colony having the sections, and so depleting the old colony, just as the young queens are hatching, that there is seldom any after-swarming. If the swarming takes place early in the season, the old colony may do something in the way of storing surplus, but, as a rule, it simply becomes a most excellent colony, with a young queen, for carrying through the winter.

As the harvest draws to a close, an extracting super is put on top of the sections, as has been already explained, or the unfinished sections may be finished up by feeding back extracted honey, or the sections nearest completion may be sold in the local market, and those not sufficiently finished for this purpose, may be extracted, and cleaned up by the bees, when they will be ready to use as "baits" to induce the bees to make an early start in the supers the following spring.

Photographed by H E. Hill.

Producing Good Extracted
Honey.

WHAT is it that gives to honey its value ? It is not simply its sweetness, which is of low power; but it is its fine flavor and rich aroma. These are the qualities which make honey what it is—a luxury—and, if we wish its use continued as a sweet sauce, we must learn to produce and care for it in such a manner as to preserve its ambrosial, palate-tickling qualities. Freshly gathered nectar is one of the most "silly" tasting and sickening of sweets. To be sure, it has the flavor of the flowers from which it was gathered; but that smooth, rich, oily, *honey* taste, that lingers in the mouth, must be *furnished by the bees.* Honey extracted when "green," and evaporated in the open air, is not only lacking in the element that comes from the secretions of the bees, but its blossom-flavor is half lost by evaporation. To be sure, evaporation must take place if left in the hive, but evaporation in the open air, and evaporation in the aroma-laden air of the hive produce different results.

One reason why comb honey is, in so many instances, found to be more delicious than the extracted, is because the former is more thoroughly ripened, and then sealed up from the air. Seldom do we find extracted honey equal to that dripping from and surrounding the section of comb honey that is being carved upon a plate. Many of those who produce extracted honey in large quantities, extracting it before it is thoroughly ripened, admit that such honey is inferior, as a table sauce, to that ripened by the bees, but they say they cannot afford to produce the best article possible. The quantity of honey is not materially lessened by thoroughly ripening it; if larger crops are secured by extracting it "green," it is the result of the stimulus given the bees by furnishing them such an abundance of empty combs. By the use of plenty of store-combs and supers, the same results, or nearly the same, may be obtained, and the ripening of the honey secured, by tiering up. The interest upon the cost of extra combs and supers is a small thing compared with the putting of unripe honey upon the market. By the use of plenty of combs, tiering them up, the work of extracting may be put off until the busy season is over. The great trouble is the lack of incentive for producing well-ripened honey for the general market. The production of extracted honey to be shipped away for some commission merchant to sell, is much like making butter to be sold at a country store. All brings the same price. White clover honey brings so much, buckwheat so much. The honey with the fine, delicate flavor, the thoroughly bee-ripened, well-preserved, superior article, will not bring one cent more in the general market than the ordinary, *pretty good* honey. Perhaps, for manufacturing purposes, there is no advantage in having such a superior article, but for table sauce there is; and the only way in which the man who produces a really superior article can hope to receive pay for his extra trouble, is by selling direct to consumers, or by establishing a reputation for his honey among dealers and their customers. The only secret in producing a superior grade of extracted honey, honey that will be the equal of that that drips from the delicate morsel of comb at the tea table, is that of leaving it on the hive until it is sealed and thoroughly ripened. Leaving the honey on the hive a few weeks after it is sealed seems to give an added ripeness or richness. Of course, robbers are ready to give trouble after the close of the season, but the use of bee escapes overcomes this difficulty. When the supers are freed from bees by the use of bee escapes, the honey is usually cold by the time it is off the hive, it having lost the heat imparted to it by the bees, and it does not extract nearly as easily as though the bees had been brushed off and the honey extracted immediately. The proper course is to stack the supers up in a warm room, one heated by a stove, until the honey is warmed through, when it may be thrown out with the greatest ease. I am aware that this system is not the one usually followed, but I believe it has decided advantages over other systems, and results in honey of a superior quality. Have plenty of combs and supers; tier up the same as in the production of comb honey; leave the combs on the hive until the honey is thoroughly ripened; remove the honey by the use of bee escapes, and warm it up artificially when ready to extract.

A Handy Uncapping Box.

This plan greatly lessens the work during the busy season, as about all there is to do is to see that plenty of surplus room is provided. If the harvest is prolonged, lasting several weeks, it is quite likely that some of the supers will be ready to come off before the harvest is over, and it may be best to remove them if they are becoming piled up too high.

It seems almost unnecessary to say that I would use a queen-excluding honey board over the brood nest. If bee escapes are to be used, the presence of brood in a super will defeat the plan, as the bees will not desert the brood. If we are going to brush off the bees and extract the honey at once, no honey that is thoroughly ripe can be successfully extracted without at the same time throwing out some of the unsealed brood if any is in the comb. With unusually deep combs in the brood nest, it may be advisable to use shallower combs in the supers, but with combs no deeper than the Langstroth, I doubt the advisability of having any shallower combs for the supers. In the production of extracted honey there is not much to choose between an eight-frame Langstroth hive and a ten-frame one, unless out-apiaries are to be established, when the ten-frame hives seem to enable the bees to bear neglect, to shift for themselves, to better advantage. They are less likely to run short of stores. Some bee-keepers use only nine combs in a ten-frame super, or seven combs in an eight-frame super, thus inducing the bees to lengthen out the cells and make the combs thicker. The honey ripens more slowly in such deep cells, but the uncapping is thereby greatly facilitated. When the combs are uncapped, the cappings should be given a thorough opportunity to drain, and, if they are kept clean, the water in which they are melted when they are rendered into wax may be made into vinegar. H. G. Sibbald, of Ontario, has the best uncapping box that it has been my good fortune to see. It is five feet long, 16 inches wide, and made in two sections, each nine or ten inches deep. The lower section is for honey, and, with the exception that the corners are halved together, it is simply a well-nailed and neatly made box, waxed inside at all joints, with a honey gate at one end to draw off the honey; the legs being a little shorter at the end having the gate, so that the honey will run off readily. The top half or section is made in the same manner, only that, instead of a board bottom, it has a wire screen bottom which allows the honey to drain from the cappings. The bottom section is halved on the inside, upper edge, and the top section halved on the outside lower edge. Being made in this manner, the lower edge of the upper section fits inside the lower one, and thus no honey runs down outside the lower box.

After the honey has been thoroughly ripened, and is extracted, and found to be in possession of all the fine qualities I have mentioned, what shall be done with it? How shall it be treated that it may *retain* its flavor? The key to success in this direction is *exclusion of the air.* Seal it up in glass jars, or tin cans, or in clean barrels; and the sooner this is done (after the particles of wax and scum have raised to the top) the less the escape of aroma. My preference is a round, jacketed, tin can, with a flat top, and a large screw-cap in the top. A five-gallon can of this kind, holding 60 pounds of honey, can be bought for about 30 cents. This style of package can be rolled on the floor. A barrel is really the cheapest package for storing or shipping honey, and when we know that honey is to be shipped to some manufactory, there is no objection to the use of barrels if they are well-made.

Upon the approach of cool weather, most honey will candy; and, if sealed up tight, and put away in a cool place, it will remain in that condition for years; and when brought *slowly* and *carefully* back to its liquid state, it will be found to have retained its original "flavor, aroma and boquet." Too much stress cannot be placed upon the care necessary in liquefying candied honey. So many think if honey does not boil it cannot be injured. The temperature of boiling water will ruin the flavor of honey. When a can of candied honey is placed over a stove, or in any other hot place, the outside of the cake of honey soon melts, and this may become very hot before the rest of the cake has dissolved.. In a tank of hot water is the best place to liquefy a can of honey, but the temperature should never go above 160 or 170 degrees; and, by the way, when melting the honey, don't loosen the screw-cap and leave it open; as it only allows the escape of the aroma.

The Marketing of Honey.

TO raise a good crop of honey cheaply, and to sell it to the best
adyantage, are two quite different processes, requiring
greatly varying qualifications. Seldom do we find all of
these qualifications in the highest degree in one person. I
believe that the majority of bee-keepers are better bee-keepers than
they are business-men; or, perhaps, salesmen is more properly the
word to use. Many of them can't get far enough away from a bee
hive to sell the honey that has been stored in it—or think they can't.
Every energy is bent to the securing of a great crop; having secured
it, many a bee-keeper is actually puzzled as to how to put it on the
market in the best shape, or how, or where, to sell.

Of course, the first step in the marketing of honey, is its preparation for the market. About all the preparation needed for comb
honey is to clean the sections of propolis, and pack them in no-drip
cases with glass fronts. If it is to be sent to a distant market, and
the shipment is less than a car load, the cases should be packed in
crates. Not boxes, as these would hide the honey, but *crates*, with
slats on the side that will allow a view of the honey. A crate may be
made to hold nine, twelve or sixteen cases. A little straw in the
bottom helps to break the force of jars. The ends of a slat on each
side, near the top of the case, are allowed to project, and thus form
handles. The position of the handles shows which side up the crate
should be kept. In fact, these handles are so *inviting* that there is
no disposition to put the crate in a wrong position. The handles are
so short that it can't be "dumped" without dumping it upon the toes
of the carriers. Cases of honey crated in this manner never tumble
over, and they reach their destination free from even the finger
marks of a dirty hand. When honey is shipped by freight, it is

quite important that the combs stand parallel with the track. If they are crosswise of the track, the bumping together of the cars breaks the combs much easier than when the combs are parallel with the track. For this reason it is well to have a large label pasted upon the top of the crate, with a large ☞ pointing lengthwise of the combs, and accompanied by the following in bold type: "Load with the hand pointing toward the end of the car, or the side of the wagon."

Much, both wise and otherwise, has been said about developing home-markets. Much depends upon the kind of home-market there is to develop, and the kind of honey there is to be sold; yes, and upon the man. The best honey producing fields are often far distant from the best markets; the best place in which to produce honey is not always the best in which to sell it. Such being the case, there is not much encouragement in trying to build up a home market, par-

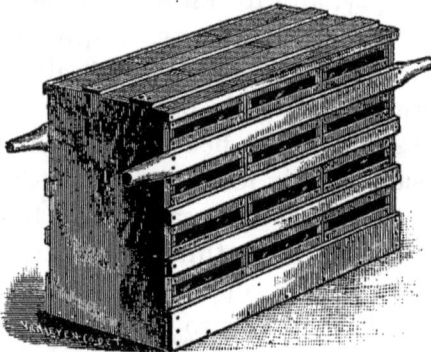

ticularly for the finer grades of comb honey, and, especially if the home market is supplied with "farmer-honey"—that raised with a lick and a brush—that is selling at retail for two-thirds what a first-class article will net when sold by a commission man in a distant city. Many bee-keepers have been able to sell to advantage, in the home markets, unfinished

sections, and lower grades of honey. In many local markets, such grades of honey will sell for as much as the choicest honey put up in "gilt edge" style, while the commission markets of a large city are a poor place in which to sell "off" grades of honey. To many grocers, in country towns, honey is honey, much the same as butter is butter. In selling honey to retail dealers, they must be visited regularly, and kept supplied with honey. In short, they must be followed up and looked after as carefully as commercial travellers look after their customers. Grocers must be educated to know that honey can't be sold unless it is kept in sight—and it should be kept under glass to protect it from flies and dust. A handsome display in a front window is a drawing card.

The putting up of extracted honey for the market calls for a large amount of thought, care and skill. Mr. McKnight, of Canada, once said that "The product of no other industry is put upon the market in such a cumbrous, uncouth and slovenly form." This may seem a little over-drawn, but it is worth thinking of. The majority of people prefer extracted honey in the liquid form, although this is largely a matter of education. There is probably no more attractive form in which it can be put up for the retail trade than in the liquid form in bottles of clear flint glass, with tin foil caps and dainty labels. A much cheaper package is that of tin, but it hides the beauty of the honey. The friction-top cans are the best tin package. They do not leak, yet they can be easily opened and the honey examined. The lack of attractiveness in the package must be made up in the label, as is the case with all goods put up in tin cans. Quite a little candied honey has been sold in *paper sacks*. The sacks are made of heavy Manila paper, paraffined, the honey put in while in the liquid state, and then allowed to granulate. The sacks can be set into small boxes, *a la* egg crate fashion, the boxes holding them square until the honey candies, when the sacks of honey can be packed for shipment like so many bricks. The purchaser can peel off the sack, and melt up the honey, if he prefers it in that state. The cost of the package is only about *one-tenth* that of tin. Every package of liquid extracted honey intended for the retail trade should have an explanatory label stating that honey will candy upon the approach of cool weather, and all packages of extracted honey, whether liquid or candied, should bear labels explaining how to liquefy the honey without injury. Right in this line, let me say that candied, extracted honey can be put up in a very attractive package. Let it candy in the square, 60-pound tin cans, or it may be bought in in these cans already candied, cut off the tin can with a pair of tinner's snips, then cut up the cube of honey into blocks of one pound each, wrap them in paraffin paper to prevent soaking, put a sheet of parchment paper of this to prevent breaking, over this slip a paper carton, and, last of all, a wrapping of white paper printed in gilt letters, raised or embossed. The A. I. Root Co., of Medina, Ohio, has been the leader in putting up honey in this "*de luxe*" style. For cutting up the honey into blocks, they use an ordinary butter cutter such as is used in the dairy trade. Thousands of pounds of honey put up in this style have been sold at retail in Cleveland at 25 cts. a pound. The beauty and novelty of the package and its contents, combined with judicious but generous advertising, made the product sell like the proverbial "hot cakes."

Many men have made large wages selling honey direct to consumers. They systematically canvass a city, or portion of a city, carrying honey with them, giving "tastes," or small samples, taking orders, and having regular days of delivery.

Of course all men are not adapted to the retailing of honey. Mr. M. A. Gill, of Colorado, who produces about two car loads annually of comb honey, says he prefers to sell it in a lump to some man who wishes to retail it, while *he* will turn his attention to the production of another crop of honey. But, even if a man does not retail his crop of honey, there is no reason why he should not use care and good judgment in selling it at wholesale. If the honey is to be sold on commission, the most important point of all is that the commission merchant be reliable. If in doubt, consult the editors of bee journals. Of course, they may sometimes make mistakes, but, usually they are quite well informed regarding the reliability of the principal dealers in honey. After all, an out and out sale of the entire crop, at the end of the season, is the most satisfactory, although so high a price is not usually realized as when the crop is sold on commission. Some bee-keepers make a business of wholesaling their own honey, that is, selling it to the same class of buyers as patronize the commission men. It requires some little time to work up such a trade, but, once it is secured, it is easily held. The first thing is to get a list of those men who use large quantities of honey. A local druggist can usually furnish the names of many of the manufacturing druggists; the groceryman can give the names of the bakers; and an advertisement in the journals will probably reach all of the bottlers of honey. These lists of names should be arranged systematically. Probably the card system would be as good a form to have them in as any that could be found. Samples of honey and prices should be mailed out to these lists, and to those who inquire for samples. Where a man has the time and ability to look after the matter, this is really a very satisfactory method of disposing of large crops of honey, year after year, at a substantial advance over what would be secured were the honey consigned to commission merchants.

Migratory Bee-Keeping.

IT is seldom that one locality abounds in all of the honey-producing plants that may be found by making short journeys in different directions. A locality unequaled for early bloom may be sadly deficient in the clover and basswood blossoming so profusely at mid-summer only a few miles distant, while a few miles farther on may be a swamp or river bottom that is of little value as a bee pasture until gorgeous with the purple and gold of autumn flowers. It will be readily seen why some bee-keepers occasionally find it profitable to move their apiaries once or twice during the season. Some notable successes in this line have been made in Florida, where the honey from the orange blossom comes in March, then a move of perhaps 50 miles allows the bees to enjoy the bloom of the saw-palmetto, and, later, another crop may be secured by moving to the mangrove region. After the harvest of sage honey is over in California, and vegetation in the mountain canons has turned dry and brown, a move of 20 or 30 miles will, in some localities, place the bees among thousands of acres of blooming bean fields from which may be gathered a white honey of fine flavor. In Canada several bee-keepers make a good profit each fall by moving their bees to buckwheat regions. In Europe bee-keepers move their bees to the heather fields, and then, later, to the buckwheat; in fact, so many move their bees to the buckwheat that a train is sometimes made up expressly for carrying the bees to these pastures. Several years ago, a younger brother of mine, who had not left home, came to my place early in August and carried home with him 20 colonies of my bees; as there was an abundance of goldenrod, boneset and willow herb in his locality, and none in mine. An empty story filled with

Willow Herb (Epilobium) in Full Bloom.

empty combs was placed over each colony, and the top covered with wire cloth. A hay rack was covered with hay to the depth of about two feet, the hives set upon the hay, and held together in a bunch by passing a rope around them. The journey of 25 miles was made without mishap. Those 20 colonies furnished 400 pounds of surplus; besides, they needed no feeding for winter, while the bees kept at home stored no surplus, and each colony required feeding, on the average, about 15 pounds.

Had buckwheat yielded well, which, in this locality, happens about once in a dozen years, nothing would have been gained by the move. The inability to foretell the honey flow in any locality, is the greatest obstacle in the way of successful migratory bee-keeping. Local showers sometimes cause a great difference in the yields of honey in localities only a few miles apart, but migratory bee-keeping does not allow us to take advantage of this; as, by the time we have moved to the locality that is furnishing honey, the flow there may be over, and, possibly, started up in the home-yard. There is nothing to be gained by changing one possibility for another of equal value. But moving to another location which promises well at a time when we *know* nothing will be gathered if the bees are kept at home, is a far different thing. For instance, only forty miles from here, on a direct line of railroad, is a locality where it is nothing unusual for 100 pounds of comb honey, per colony, to be secured, yet *nothing* is in bloom here at that time. The expense of moving to and from a locality no farther away than this need not be so very great. From 30 to 40 colonies can be moved on a hay rack; or a special rack might be made which would accommodate 50 colonies. An apiarist who is going to practice moving his bees to secure better pasturage, must have hives, fixtures, and other arrangements suitable for that purpose. The arrangements ought to be such that three or four minutes would be sufficient for preparing a hive for moving. One of the greatest advantages of fixed, or self-spaced, frames is that they need no fastening when the apiary is to be moved. Of course bees moved in hot weather must have abundant ventilation; but this alone will not save the *brood*, if they are long confined. To save the brood the bees must have plenty of water.

Some localities are blessed with an almost continuous flow; spring flowers, white clover, basswood, and fall flowers; and, by the way, a man who is to make a specialty of bee-keeping ought to seek such a locality; but many who are already engaged in bee-keeping are permanently located, have friends and relatives living near, and prefer not to seek a new location even if the profits would be thereby increased. Then, again, it is difficult to find a first-class locality for

Moving to Better Pastures.

clover and basswood that is equally good for fall flowers; and, the better the locality the greater the danger of its being overstocked by its very attractiveness bringing together so many bee-keepers.

There is no question but what many bee-keepers can secure a bountiful crop of fall honey by moving their bees at the right time, but a word of caution may not be out of place right here. Some fall honey, that from aster, for instance, is sadly unfit for winter stores. So disastrous has fall honey proved for winter stores, in some localities, that the bee-keepers there have given up trying to winter their bees unless they substituted early gathered stores, or fed sugar. I know of one bee-keeper in such a locality who secured bountiful

On the Road.

crops of fall honey from the surrounding swamps, but was utterly unable to winter his bees, prepare them as he might, and he finally fell to shaking them off the combs at the close of the season (thus saving the honey), and restocking his apiary in the spring with bees from the South. So, I say, beware when you move your bees to fall pastures of asters and swamp flowers.

There is another form of migratory bee-keeping that has long been the dream of apiarists, that of starting with an apiary in the South at the opening of the honey season, and moving northward with the season, keeping pace with the advancing bloom, thus keeping the bees "in clover" the entire summer. The difficulties to be overcome are largely those of transportation. There is no single line of railroad running north and south for a sufficiently long distance to make a success of migratory bee-keeping. When shipping bees by freight, on the migratory plan, the delays at junction points are not only vexatious but disastrous. It is for this reason that longing eyes have been cast at the Mississippi river and her steamboats, and, once, Mr. C. O. Perrine tried moving several hundred colonies up the Mississippi on a barge towed by a tug. The plan was to run up the river nights, and "tie up" during the day to allow the bees to work. There are several reasons why the plan was a failure. The start was made too late in the season, and accidents to the machinery of the tug caused delays. In order to overtake the bloom it became necessary to confine the bees and run day and night. The confinement for so long was very disastrous to the bees. Those who aided in the enterprise believe that, rightly managed, the plan might be made a success. Mr. Byron Walker, who has had much experience in moving bees from the South, greatly favors the Mississippi plan of migratory bee-keeping. He would not put the bees on a barge and tow them with a tug, but would load them upon a regular steamer running up the river, setting them off at some desirable point, and then shipping them by another boat to another point farther up the river, as the flow began to wane. In the fall he would take the bees back South for the winter.

Right here a hypothetical question comes to mind. Supposing that an apiary moving up the Mississippi secures as much as six ordinary crops of honey—six times as much as a stationary apiary— would this be more profitable than six stationary apiaries? In other words, which is the more promising field for enterprise, following up the season, or establishing out-apiaries? Upon this point there are many things to be considered, and varying circumstances would lead to different decisions. To establish six apiaries would require considerable capital, and the labor of caring for the honey crop would all come at one time, while there would be only one "chance" of securing a crop. With the migratory plan, only one apiary would be needed, and the work of caring for the surplus would not come all at the same time. With the stationary apiaries there would be no expense for transportation, which is a big item.

Out-Apiaries.

WHEN a man starts an out-apiary it is because he thinks his home-yard overstocked; that he will get enough more honey for the division to pay for the extra labor incurred. Overstocking is one of the most puzzling questions connected with bee culture. We all know that a locality *can* be overstocked; but localities, seasons and bee-pasture are so variable that it is impossible to lay down any set rules in regard to the number of colonies needed to overstock a locality. It must not be forgotten that the yield per colony, yes, and in the aggregate, may be diminished to considerable extent by overstocking ere the establishment of an out-apiary would be a profitable move. At times of great honey flows it is probably practically impossible to overstock a locality; the overstocking occurs during the lighter yields. There is occasionally a man, notably Mr. E. W. Alexander, of New York, who makes a success of keeping a very large number of colonies in one apiary, by feeding during times of scarcity. Mr. Alexander has secured as high as 75 pounds of extracted honey per colony from 700 colonies in one yard. This question of how many colonies will justify the starting of an out-apiary is one that must be settled according to the circumstances of each individual case, and can never be decided with more than approximate correctness.

I have had no experience with out-apiaries, but I believe that the majority of the inexperienced have erroneous ideas as to the difficulties and expenses attending the establishing and management of out-apiaries. Land must be bought or hired; some sort of a building or shelter secured; and a conveyance of some kind will be needed for carrying bees, tools and supplies. Then, in the Northern States,

Part of a 700-Colony Apiary Belonging to E. W. Alexander, Delanson, N. Y.

there is the preparation of a cellar for wintering the bees, or they must be carted home in the fall, and back in the spring, or else protected upon their summer stands. But when a man begins to number his colonies by the hundreds, he knows that *something* must be done. Even if out-apiaries are not so profitable as home-apiaries, they are not usually run at a loss, while the removal of the surplus bees at the home-yard, allows that to make better returns.

When it is finally decided to start an out-apiary, how far away shall it be located ? We have been repeatedly told that, ordinarily, three miles mark the limits of a bee's foraging grounds; hence, if apiaries were placed six miles apart, there should be no encroachment. But it must be remembered that the pasture ground of each apiary is somewhat circular in form, hence they might be moved towards each other to considerable extent without one encroaching very much upon the other. Dr. Miller has given a very happy illustration: Lay two silver dollars side by side. Lift the edge of one and slide it over the edge of the other. Notice how far it may be pushed over without covering a very large portion of the other. Notwithstanding all this, those who have had experience in the matter are not inclined to place out-apiaries nearer together than four miles, and prefer to have them five or six miles apart. When the team is hitched up and on the road, a mile or two more travel does not take so *very* much time, and the increased yield may more than make it up. We cannot always secure the exact spot desired for the location of an out-apiary, and it would probably be well to go a little farther than really necessary, rather than to crowd some other apiary.

The mode of travel to and from out-apiaries will depend upon circumstances. Some men have a honey house, with extractor and kit of tools at each apiary; and ride a bicycle to and from the work, storing the honey at or near the apiary, and hauling it home at their leisure. A few men have been fortunate enough to be able to locate out-apiaries near some trolley line by means of which they can go and come any hour of the day. Probably the majority find horses the most desirable means of travel; in which case one set of tools will answer for several apiaries; it is even possible to dispense with honey houses at the apiaries, a tent being carried, and slipped over a light frame-work kept standing at each yard. A covered wagon is sometimes made to answer as an extracting room.

After locating an out-apiary, and deciding upon the mode of travel to and from it, the matter of management brings up several

Covered Wagon for use as Extracting Room at Out-Apiaries.

questions. Shall comb honey be produced, or shall the honey be taken in the extracted form ? Shall it be managed upon the visiting plan, or shall a man be kept there during swarming time ? I believe that, in the majority of cases, extracted honey is produced in out-apiaries; as by this plan, swarming can be nearly controlled, and the apiaries visited only at intervals. Mr. E. D. Townsend, of Michigan, has successfully managed an apiary for extracted honey by visiting it only four times a year. The bees were in ten-frame, Langstroth hives. At the approach of the white clover flow he visited them to remove the packing and put on two upper stories of combs. He visited them twice to extract and again to pack them up for winter. His profits averaged $150 for each visit. He approves of visiting an apiary oftener than this, but his experience shows what can be done. The reason for not visiting this apiary oftener was that it was 50 miles from home. And this brings up another point in connection with out-apiaries: If they are widely scattered, with varying kinds of pasturage, there is almost a certainty of securing a crop each year from *some* of them.

The difficulty in the past in managing out-apiaries for comb honey has been that of controlling swarming, but the discovery of "shook-swarming" changed all this, and gave a wonderful impetus to the establishment of out-apiaries. By visiting an apiary once a week, and "shaking" every colony that has started queen cells, there will be little, if any, swarming. A few bee-keepers succeed in preventing swarming by removing the queens, at the beginning of the swarming season, but the practice has never been generally adopted.

As many colonies ought to be placed in an out-apiary, as the location will bear; certainly enough to make a day's work at each visit during the busy season, as it would be unprofitable to drive off five or six miles to do only part of a day's work.

In those parts of the country where out-door wintering is uniformly successful there need be no question as to how bees shall be wintered at an out-apiary, but where cellar-wintering must be de-depended upon, a choice must be made between building a cellar at each apiary, and that of carting the bees home in the fall, and out again in the spring. If the bee-keeper knows, positively, that an apiary is permanently located, it may be worth while to consider the construction of a cellar on the ground; but, usually, there is more or less shifting about of out-apiaries, and, unless *too* far from home, I should be inclined to follow Mr. P. H. Elwood of New York in bringing them home in the fall and carrying them out in the spring. Mr. Elwood sometimes has many as 1,000 colonies in one cellar.

Out-Apiary in Raspberry Region of Northern Michigan.

The logs "banked" just beyond the apiary are hard wood timber, and where this is cut off the wild red raspberries spring up and furnish great quantities of nectar.

Mr. E. D. Townsend, whose out-apiaries are widely scattered, buries his bees or puts them in "clamps," as it is called; and where the soil and location are suitable this is an excellent method of wintering bees.

Lake of the Woods, in the Raspberry Region of Northern Michigan.

Perhaps few can understand the longing there is in the heart of the author of this book to build himself a real log cabin, with stone fireplace and chimney, on the shore of some one of the beautiful little Inland lakes of Northern Michigan, establish an apiary hard by, right in the woods, and pass at least a portion of each summer in that sylvan retreat. What a place to take bee-keeping friends in the autumn, when the evenings could be spent around a fire of blazing pine knots in the fireplace.

House Apiaries.

A HOUSE apiary, as indicated by its name, is an apiary kept in a house, the bees passing out through openings in the walls. Formerly, the hives were built, permanently, in the house; the shelf upon which they set forming the bottoms, the walls of the building forming one side, and each division board between any two colonies forming one wall for both colonies. Eventually it was discovered that building the hives into the building in this stationary manner curtailed or complicated many of the manipulations. For instance, if a colony swarmed, and it was desirable to hive the swarm upon the old stand, moving the parent colony to a new stand, it could be accomplished only by removing the combs one by one, and carrying them to a new location. When the ordinary hives are used, any colony can be picked up and carried to any location. A swarm can be hived out of doors, then the hive picked up and carried into the house. Still another point: Some bee-keepers like a house apiary for summer, but find it a very poor place in which to winter bees, hence they build a cellar under the house, and winter the bees in the latter, this course being possible only when the hives are movable.

It will be seen that although we have a house apiary, we also need the regular hives, just the same as though they were to be kept out of doors, with this exception, that if they are to be used exclusively in the house, they may be made of cheap lumber and left unpainted. The same may be said of the supers or upper stories. If we must have regular hives, why have a house apiary? Well, here are some of the advantages: The house can be locked against thieves; the colonies, apiarist and his tools are brought close together, and

House Apiary of Fred H. Loucks, Lowville, N. Y.

under shelter; and this latter point is very important, especially in the management of a series of out apiaries that are to be visited periodically. Rain puts an end to bee-work in the open air, and three or four days of rainy weather sadly demoralizes the plan of visiting an apiary once a week, when there is an apiary for each day in the week. In a house apiary the work can be continued regardless of the rain. Of course, there would be the travelling to and fro in the rain, but rubber coats and blankets overcome that difficulty. Shelter from the hot sun is often a great comfort. In taking off honey there is never any trouble from robber bees. Bees are more peaceable, that is, less inclined to sting, when handled in a house. In

General View of Mr. Ludington's House Apiaries.
(Honey House in the Center—Shop in the Background)

short, the advantages, with one exception, are nearly all with the house apiary, and this exception is the cost of the building. Formerly there was the objection that the removal of the surplus liberated many bees inside the building, where they were a great annoyance upon the windows and under foot. The introduction of the bee-escape has removed this most serious objection. By means of the escapes surplus can be removed with scarcely a bee entering the building, and these few find their way out through the escapes with which the doors and windows are provided at the top.

Probably the only really serious objection to the house apiary, aside from its cost, is the great likelihood of queens being lost while on their wedding flight; that is, of their entering the wrong hive upon their return. The trouble arises from the great number, similarity and regularity of the entrances. To help to overcome this difficulty, different portions of the house are often painted different colors, and different designs are placed about the entrances. Some bee-keepers have found it desirable to rear their queens outside of the house and introduce them when needed.

Mr. A. A. Ludington, of Verona Mills, Michigan, uses small house apiaries, made of cheap lumber, and winters his bees in a cellar. Instead of setting his hives upon shelves, he hangs them up against the walls by means of heavy wire loops. The bottoms to his hives are hinged so that they can be let down. This allows of an easy examination of the lower edges of the brood combs where the bees are almost certain to build queen cells if preparing for swarming, thus he is able to foretell swarming very quickly without so much as opening a hive. The bees can easily be driven up among the combs by the means of smoke, when, by using a hand mirror, if necessary, a view can be obtained that extends up quite a distance between the combs. If the light is insufficient, some one can stand out of doors with another mirror, and throw a flood of sunshine under the hive that is being examined.

Foul Brood.

FOUL brood is a bacterial disease of the larvae or brood of bees. Once a single spore of the disease comes in contact with a larva, or is fed to it, it begins to increase with wonderful rapidity; the bacteria feeding upon the larva as maggots feed upon the carcass of a dead animal. The larva soon dies and turns a dull brown, something about the color of coffee after milk has been added and it is ready for drinking. The dead larvae soon lose their shape, and settle down into ropy, gluey masses having an odor somewhat similar to a poor quality of glue when it is warming on the stove, being made ready for use. In the earlier stages this odor is seldom noticeable, but, as the disease increases, this odor becomes quite pronounced. If a match, or a wooden tooth-pick, or something of this nature, be thrust into a dead larva, and then withdrawn, the dead matter will adhere to the stick, and draw out in a ropy string, perhaps an inch in length, when it will break and fly back. The dead larva finally dries down into a thin brown scale upon the lower side of the cell. A large share of the larvae reaches that stage where the bees seal it over, but, for some reason, the cappings often become sunken, and sometimes contain holes. Of course, the healthy brood hatches, while the diseased brood does not, and soon the combs present a peculiar, speckled appearance from part of the cells being empty, while others are sealed with dark, ragged cappings. When the bees attempt to rear another larva in a cell where a larva has died of foul brood, it is certain to be a failure. This larva, too, dies of the disease. If honey is stored in the cell it becomes contaminated with the germs of the disease; and if fed to larvae infects them with the disease. The combs finally become so contaminated with the disease that scarcely any brood can be reared. The old bees die off from natural causes, and, their being no young bees reared to take their places, the colony dwindles away until it becomes a prey to robber bees who carry home the honey, and thus start the infection in their own hives. In this way the disease is spread from hive to hive, and from apiary to apiary.

Such, in brief, is foul brood; and, as there is no apiary in which there is not a possibility that it may appear, every bee-keeper ought to be able to distinguish it, and to know what to do when he is so unfortunate as to find it in his apiary. From reading the published descriptions, many bee-keepers have formed exaggerated ideas regarding the appearance of foul brood, especially of its appearance in its *first* stages. They are looking for combs black with slime and rottenness, a stench strong enough to knock a man down, and colonies dwindled away to mere handfuls. The possession of these exaggerated ideas by bee-keepers has allowed foul brood to gain a strong foothold in many an apiary long before the unfortunate owner ever dreamed of its presence. At first, only a few diseased cells will be found. Of course, it is not advisable that a bee-keeper be continually opening brood nests, and critically examining combs for foul brood, but there are certain danger signals that it is well to bear in mind. If a colony shows signs of listlessness; or many dead bees are found in front of the hive; or, if a peculiar, unpleasant odor is noticed, it would be wise to make an examination. *Whenever* handling combs of brood, it is well to glance *understandingly* at the brood. Notice, if the "pearly field" of unsealed larvae is unbroken. If there are desolate patches; and the sealed brood is scattering and in patches instead of in solid sheets, examine more critically. If some of the larvae are discolored, shapeless, ropy, ill-smelling, some of the cappings sunken, perhaps perforated, foul brood is present. The one *sure* symtom of foul brood is the ropiness of the larvae. If a splinter be thrust into a dead larva, and withdrawn, the matter will adhere to the splinter, and "string out," perhaps an inch, or more, then break, and the two ends fly back to the points of attachment. Without this viscidity there is no foul brood—with it there is always foul brood.

Right here it might be well to say that all dead brood found in the combs is not foul brood. There is chilled brood, starved or neglected brood, "pickled" brood that comes and goes from what cause no one yet knows, but in all of these the ropiness is lacking. In the majority of cases the outer skin of the larva does not seem to decay, and enables the operator to draw the whole larva from the

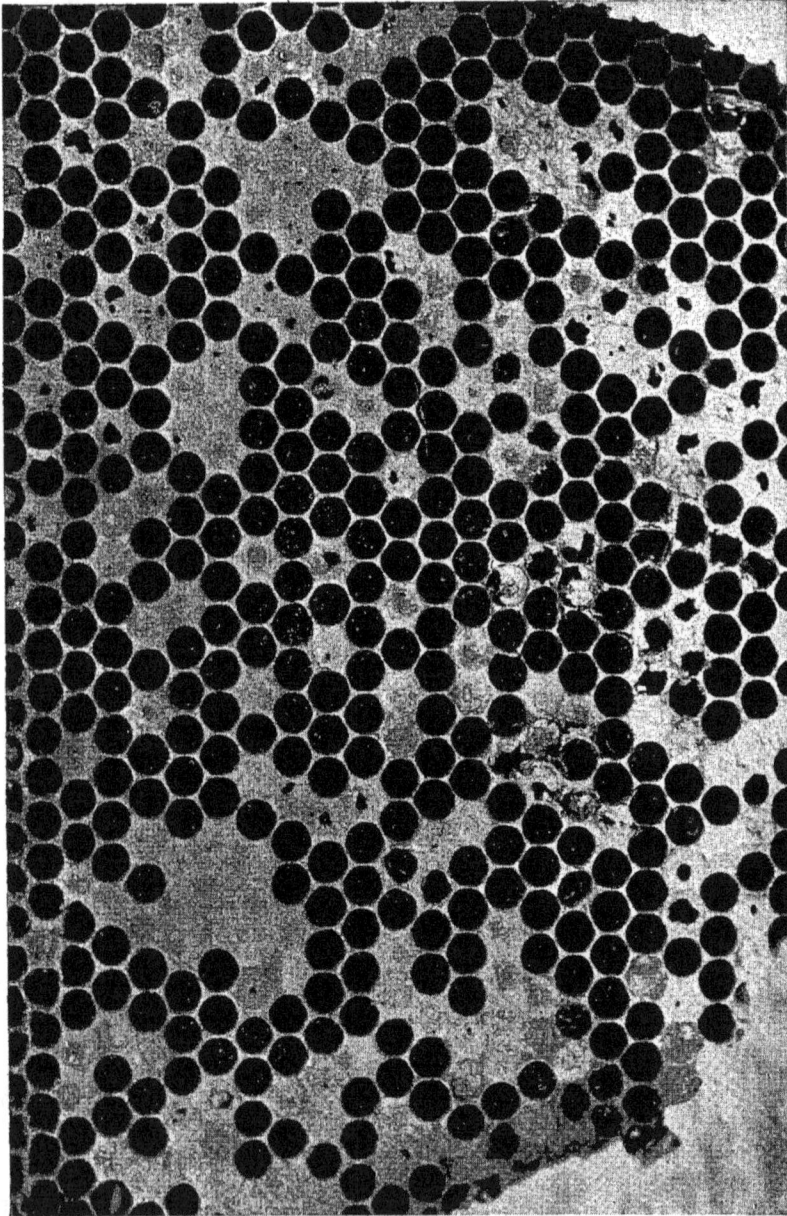

An Advanced Stage of Foul Brood.

cell. Then there is black brood, that has caused so much havoc in New York. In this the dead larva is more of a gelatinous nature than anything else. It may sometimes string out quarter of an inch, but never more than that, while foul brood will string out at least an inch, and sometimes much further. Black brood turns slightly yellow, then a dark brown, and finally becomes black, hence the name. It does not emit that gluey or "old" smell that comes from foul brood. There is scarcely any odor, and what little there is might be called a sour or fermenting smell, like that from decaying fruit. Black brood is very similar to foul brood. It spreads in the same manner, and treatment is the same as that for foul brood.

To come back to foul brood once more. The symptoms enumerated above will be seen only during the breeding season. In a strong colony, after the breeding season is over, the cappings are all cleared away, and the dead brood is entirely dried up—mere scales almost the color of old comb itself, lying fast to the lower sides of the cells, and drawn back more or less from the mouths of the cells. There is probably no symptom of foul brood that is more difficult for the novice to detect than these dried down scales, and, as just explained, except in the breeding season, they are the only evidence that can be found of the disease. Here are the instructions given by Mr. N. E. France, Inspector for Wisconsin, for finding these scales: "Bring a brood comb up from the hive to the level of your chin; then tip the top of the comb towards you, so your view strikes the lower side-walls (not the bottom) of the brood-cells about one-third distant from the front end of the cells. Then turn so that the rays of bright light will come over your shoulder and shine where your eye is looking. Gas or electric light will not take the place of good daylight. On the lower side-wall, a little back from the front end of the infected cell, will be seen the dead larva bee, nearly black, with a sharp pointed head, often turned up a little, the back portion of the bee flattened to a mere lining of the cell, often no thicker than the wax in the wall of the comb. The base, or bottom of the cell, likely looks clean; also all of the other side-walls of the cell."

Honey is the means by which the disease is usually carried from one hive to another. Mr. Frank Cheshire says that the mature bees, the queen, and even the eggs, are infected in a diseased colony. Be this as it may, where the bees of an infected colony swarm, or are shaken from their combs into a new or clean hive, and given no combs in which they can store the infected honey that they have brought with them, the brood hatched afterwards, in this newly formed colony, remains free from disease. Foul brood is often brought into an apiary by the bees robbing some diseased colony in

FOUL BROOD

Looking for the Dried Down Scales of Foul Brood.

The white line shows the angle of vision, and at which the light should fall.

a neighboring apiary, and bringing home the honey. The buying of second-hand honey cans often brings foul brood into an apiary. If the bees gain access to them they soon lick up any honey that may have dripped upon the outside of the cans; or the bee-keeper may rinse out the cans and throw out the water upon the ground where the bees will come and suck it up. I have known a bee-keeper to clean out a lot of second-hand cans, and feed the honey directly to the bees, with the result that foul brood developed in every colony that was fed. In rare instances the buying of queens from a distance has introduced foul brood into an apiary. The queens themselves had nothing to do with disseminating the disease, but the bees and honey that accompanied them brought with them the germs of the disease. It is a safe plan to put the new queen into a clean cage and destroy the accompanying bees and cage. After foul brood is once introduced into the apiary, it is disseminated by robbing, by the careless exposure of infected honey, by changing combs from hive to hive, or by extracting honey from infected combs, thus contaminating the extractor and other combs that may be brought in contact with it.

When foul brood is discovered in an apiary, what shall be done? In the first place don't "lose your head," as the saying is. Don't be in such a haste to be rid of the pest that a crop of honey is lost, and the work of eradication imperfectly performed. Curative operations can be carried on only during a successful honey flow, when bees will not rob. If foul brood is discovered after the honey season is over, treatment must be postponed until the following year.

The entrances of all weak colonies should be contracted, and any colony too weak to make the proper defense, or so weak that it is not likely to pass the winter, better be destroyed at once.

The spraying of the combs with acids, the fumigating of them with formalin gas, the feeding of the bees with medicated honey, are all of little avail so far as eradicating the disease is concerned, but may do much in the way of checking the disease and preventing its spread. By the proper feeding of medicated syrup in the spring, the disease may be so held in check as to interfere little, if any, with securing a crop of honey. This medicated syrup is made by mixing one ounce of salicylic acid with sufficient alcohol to dissolve it, after which it may be stirred into about 25 quarts of a not too thick syrup or honey. We should begin feeding the bees this syrup as early in the spring as they will take it, keeping each diseased colony supplied with syrup until the flowers yield fairly well. Weak colonies better be united, but there must be caution in doing the work, gradually bringing them together, that the bees may not be scattered into other hives, until they are side by side before the union is made.

Finally, the main honey flow comes on. With us this is the last of May, or fore part of June. Now is *the* time for treating the diseased colonies. Any colony that is strong, and almost or nearly in a condition to cast a swarm, may be treated as follows: Set the colony just back of its old stand, and upon the stand place a hive the frames of which are furnished either with full sheets of comb foundation, or with starters of the same. Remove the combs from the old hive and shake off most of the bees, in front of the new hive. Nothing more need be done to the colony in the new hive. Ere it can rear brood it will have consumed any infected honey that the bees may have brought with them. Don't use drawn combs instead of starters or foundation, because the bees might store some of the infected honey in the comb, where it might remain until brood was being reared, when, if this honey should be fed to the brood, the disease would be again started. I have never found it necessary to give the bees a second set of frames, and a second shaking, as is practiced by some. Neither have I found it necessary to boil or otherwise disinfect the hives. The old hive, with the combs of brood, is placed upon a new stand. Sometimes two sets of combs from which the bees have been shaken are united. In ten days a young queen, or a ripe queen cell, may be given the old colony. In 21 days from the time the bees were shaken off, just as all of the healthy brood has hatched, and the young queen is beginning to lay, the colony may be again treated exactly as it was at the first shaking, when the result will be another healthy colony, while the old combs will be entirely free from brood, and should be taken to some place of safety (where no bees can gain access to them) and eventually treated as may seem best.

Colonies not populous enough to make a good colony, each, when shaken, may be treated in "pairs." We select the first pair, set one of them aside, as was done with the strong colony, and put a hive containing frames furnished with foundation, in its place. We now shake out the bees into the new hive, as before, only we get *all* of the bees, as well as the queen. We now put the old hive with the brood on the stand of the other hive of the "pair," bringing the latter to the location where the first "shaking" took place, and shake out the bees and queen in front of the hive into which the bees from the first hive were shaken, the combs of brood being taken back to their old location and united with the combs of brood from the first-shaken colony. We thus get only one "shook swarm" from two colonies, but it is stronger for that reason. The united colonies of brood will be given a young queen in 10 days, and then shaken upon a new set of frames in 21 days, as was done with the populous colony first described.

A good part of the success of this plan is owing to the medicated food given in the fore part of the season. Of course, the same treatment may be given without it, and will be equally effective, so far as a cure is concerned, but the condition of the colonies, and the amount of surplus secured, will be far different.

There is still another method of treating foul broody colonies in which there is no shaking off of the bees; and it has always been a wonder to me that it has not come into more general use. The plan originated with Mr. M. M. Baldridge, of St. Charles, Illinois, and is called the Baldridge method. It is based upon the fact that when a bee leaves a hive naturally, in quest of honey, its sac is free from honey, and it might enter a healthy colony without infecting it with disease. Of course, when it returns with a load of newly gathered nectar, it is still in that harmless condition. Here is the method of management: Bore a hole in the front of the hive, just above the entrance, and near the side of the hive. Over this opening fasten a bee escape in such a position that bees can pass out of the hive through the escape, but can not return. Next cage the queen of the colony, laying the cage on top of the frames. The following morning go to some healthy colony and select a comb of sealed brood, either with or without the adhering bees, place it in an empty hive, filling out the hive with frames filled with foundation, and set the hive thus prepared upon the stand of the diseased colony, setting the latter to one side, so that the two hives will stand side by side, close together, and fronting in the same direction. Have the bee escape as near as possible to the entrance of the new hive that is on the old stand. It will be seen that all flying bees will return and enter the new hive on the old stand; and, as fast as the bees leave the old hive by means of the escape, they will return and join the newly formed colony upon the old stand, as it will be impossible for them to enter the old hive. At sundown of the first day after setting the old hive upon a new stand, open the hive carefully, take away the caged queen, being careful to take no bees with her, and let her run into the entrance of the new hive. All of this work of closing the entrance of the old hive, setting it upon a new stand, and removing the caged queen, should be done as carefully as possible, so as not to disturb the bees and induce them to fill themselves with honey. Nothing more need be done for about a month, by which time the brood will all have hatched, and the bees have left the hive and joined the new colony. The hive should be opened in some close room from which no bees can escape; and, should a few stragglers remain in the hive, they should be destroyed. The combs are now free from bees and healthy brood, and ready to be treated as seems best, while there is a healthy colony in the apiary where once stood the one diseased with foul brood.

When freed from bees and healthy brood, no matter what the method employed, the combs may be emptied of honey with the extractor, and then melted into wax. Of course, an extractor thus used must be most thoroughly cleansed before it is again used for extracting combs of honey from healthy colonies. For disinfecting the extractor I would use a strong solution of salicylic acid, pouring it on boiling hot from the spout of a tea kettle. The matter of cleaning the extractor is one about which one cannot be too thorough. Honey from such combs ought not to be placed upon the general market, as consumers are liable to throw out an empty package where neighboring bees will come and clean it up. Some bee-keepers ship such honey to bakers where the heat used in baking will destroy any germs that may be in the honey. Thorough boiling of the honey will kill the germs and make it safe for use in feeding the bees, but before the honey is boiled it must be mixed with an equal quantity of water. Some advise boiling the hives, or burning them out on the inside by painting them over with kerosene and setting it on fire, but I have seen so many hives used without taking any such precautions, that I have come to doubt their necessity. Mr. McEvoy, Inspector of apiaries for Ontario, says that he has cured thousands of cases of foul brood without any such disinfecting, and considers it wholly unnecessary. Some have advocated the burning of the combs with no attempts at saving the honey and wax. If only a few colonies are to be treated, this might be advisable, but the owner of a large apiary quite generally affected with foul brood, can well afford to take the necessary precautions whereby the honey and wax may be saved. Whoever undertakes such a job must remember, however, that eternal vigilance is the price of success. *One drop* of the infected honey secured by a robber bee, means disease once more in the hive to which it is carried. It has been suggested that the extracting, etc., be done down cellar. It is a cool place in which to work, and the bees can the more easily be kept out. Others have melted up the combs at night when no bees were flying. There is really no necessity of destroying the combs and the honey they contain. If a man can not, or will not, exercise sufficient care, it would certainly be better to burn them; but, if he has "gumption" enough to succeed as a bee-keeper, he can save the combs from destruction. It would be well, however, for all to bear in mind that one little "forget" may compel a repetition of the whole business.

Of late, the bee-keepers in some parts of Colorado have been following a plan whereby a man may keep his apiary fairly free from foul brood, even though located in a foul broody district. It is well-known that shaking the bees of a foul broody colony into a clean hive, and allowing them to build a new set of combs, frees them from the infection; well, these Western men, just at swarming-time, treat *every colony* in the apiary in this manner—make a wholesale sweep of the matter. As this is done at a season when the honey flow has commenced, and there will be no more robbing until the season is over, the apiary remains free of the disease for that season. It is simply forced swarming on a wholesale scale. The old hives are given new locations, and perhaps the combs of two or more colonies are put together upon one stand. Perhaps it ought to be mentioned that the combs are not shaken entirely free of bees, some being left to care for the unsealed brood; and caution is exercised that the work be not done too early in the season when there would be danger of chilled brood or from robbers. At the end of three weeks, the brood has all hatched, when the combs are shaken entirely free of bees, the latter, of course, going back into the hives and building new combs, thus establishing colonies that are free from the disease. The honey is then extracted from the old combs, and the latter rendered into wax. It is asserted that the wax will pay for the labor, while the new combs are built *at a profit*. I believe that this plan can be successfully followed in the East as in the West; although, of course, the Western harvest is much longer than ours.

RAVAGES OF THE BEE MOTH'S LARVAE.

Apiarian Exhibits at Fairs.

FOR 15 years I did not miss making an annual exhibit of bees and honey at our State fair; and, for three or four years, I also made exhibits at the State fairs of Indiana, Wisconsin, Illinois and Missouri. It will not pay to travel from State to State with an exhibit, unless the exhibit is unusually large and attractive—enough so as to win the lion's share of the premiums. On the other hand, it will not pay to get up a large, expensive exhibit, unless it is to be exhibited at several fairs. In order to thus make a "circuit" of several State fairs, it is necessary to charter a freight car, and travel with the exhibit. In no other way is it possible to avoid fatal delays at transfer points. The work is terribly hard; there is the packing up at night, and travelling nights in a freight car, the "hurrah boys" of getting upon the grounds and the exhibit set up in time, and the friendly rivalry with competitors, but there is a fascination about it that, to an old exhibitor, is almost irresistible.

There has been, in times past, some opposition to these apiarian exhibits, on the ground that they were often made by supply dealers who, in their eagerness to do business, did not hesitate to urge a man to become a bee-keeper, in order to effect a sale. If the fruit of the seed sown at these gatherings *were* a crop of producers, I might admit that, possibly, there would be some injury to existing bee-keepers, but, after the experience that I have had, I am thoroughly convinced that nothing of the kind occurs; in fact, the exhibition of hives, implements, and large quantities of honey tastily put up, impresses the crowd with the true importance, magnitude and complexity of modern bee-culture; imparting the idea that the bee business is quite a business—one that cannot be picked up and learned in a day by some Tom, Dick or Harry.

Anything that increases the consumption of honey is a benefit to the pursuit; and, as usually managed, these bee and honey shows call the attention of crowds of people to the excellence and deliciousness of honey as a food; and the producer and consumer are brought face to face. At a fair, people are abroad with a disposition for sight-seeing, investigation, and the purchase of novelties and nick-nacks; and, a fine display of honey, together with its sale in fancy packages, can not help benefiting the exhibitor as well as the pursuit. Honey to be sold at fairs ought to be put up in small packages. It may be difficult to put it up in packages so small as to be sold at five cents each, but I believe it has been done, while there is no difficulty in putting honey in packages that may be sold for 10 or 25 cents each. People at fairs don't wish to be burdened with heavy or bulky packages, and the honey must be put up in such shape that it can be eaten on the grounds, or else carried in the pocket or hand bag with no danger of leakage. I remember that, one year, at the Michigan State fair, Mr. H. D. Cutting sold nearly $40 worth of honey put up in pound and half-pound, square, glass bottles and in small glass pails. One year, at the Detroit Exposition, at least 1,500 pounds of "honey jumbles" were sold at a cent apiece, by three exhibitors in the bee and honey department. These "jumbles" are made with honey instead of sugar, and, for this reason, retain the desired amount of moisture for a long time. In selling them at a fair, a box of them is opened, placed on the counter, and tipped slightly outwards, so that visitors can easily look into it. The cakes are round, with a hole in the middle, and the upper side is of a golden yellow, with a sort of granular appearance that is very inviting. This side of the cakes is turned uppermost. Paper sacks are filled with cakes, putting five in a sack, and a neat placard announces: "Honey Jumbles; Made with Honey Instead of Sugar. Five in a Sack and Five Cents a Sack." Another thing that may be sold at an apiarian exhibit with even greater profit than the honey jumbles, is honey lemonade—if the weather is hot, if it isn't, there is no use of attempting its sale. Here is the way to make it: Into 12 quarts of water squeeze the juice of a dozen lemons, add two pounds of basswood honey and a teacupful of sugar. Basswood honey being of such a strong flavor, gives more of a honey flavor. Keep the lemonade cool with ice in some large vessel. I used a stone ware churn. Keep on the counter a glass pitcher filled with lemonade, putting in small pieces of ice, also a few slices of lemon. Then have a placard

Comb Honey Stored by one Colony at the Pan American Exposition.

read: "Honey Lemonade: Most Delicious Drink on the Grounds; Only Five Cents a Glass." I have sold as high as $20 worth of this in one hot afternoon, and the profits are at least three-fourths. Such exhibitions and sales certainly do the pursuit of bee-keeping no harm, while they bring a profit to the exhibitor.

Neither ought the social feature to be overlooked. Every bee-keeper attending the fair hunts up the "Bee and Honey Department," and only one who has been at an exhibition knows of the many new acquaintances thus formed, and the old ones that are renewed. It is well to have one day set apart as "Bee-Keepers' Day," giving the date in advance in all of the bee journals, then all bee-keepers will be present on the same day. When possible to do so, it is an excellent plan for two, or three, or more, exhibitors to club together and take a tent, or a portable house, each bringing his share of bedding, provisions and utensils, and live *a la* picnic during the fair. Some of the happiest hours of my life have been spent in going through just such experiences with boon companions.

I doubt very much if the exhibition of bees at fairs is any great advantage to the pursuit. The most that can be said in its favor is that they attract attention. There is certainly no necessity of exhibiting full colonies, unless it might be at some permanent exhibition that is to last several months, when the bees can be allowed to fly, *a la* house apiary, provided the apiarian department is on the second floor. A single-comb nucleus with a queen and a few drones and workers, together with brood in different stages of development, can be made to show more that is really interesting than can be shown with a full colony.

Of course it is impossible to go on and cover, in detail, all the points in regard to planning and putting up an apiarian display, as circumstances vary greatly, but here are a few hints: Extracted honey should be shown in glass. Not common green glass, but in white, flint glass. Have tin foil over the corks, and *small* tasty labels. Aim to get a white, or light colored background for extracted honey. A dark color gives it a dull, or muddy, appearance. I know of nothing better, or more appropriate, for this purpose, or as a background for any apiarian display, than honey producing plants pressed and mounted on white card boards and the cards tacked upon the wall back of the exhibit. A pyramid of extracted honey in bottles, in front of a window, is a beautiful sight; the light "shimmering and glimmering," as it passes through the bottles and their contents. Comb honey must be in cases with glass next the comb. For several years, I exhibited honey built up into a circular pyramid. First there was made a stout, board wheel, perhaps eight feet in diameter.   This was placed perhaps two feet from the floor, being supported by blocks or boxes.   Attached to the edge of this wheel, and hanging down, for all the world, like a woman's skirt, was a sort of valance made of thin, blue cambric ornamented with some neat design of gilt paper fastened on with paste.   Around the edge of the wheel, upon its upper surface, was set a row of shipping cases of comb honey, with glass sides turned out.   On top of this row was set another row, the cases of this row "breaking joints" with the one below.   Perhaps four rows were placed in this manner,

Honey Exhibit of the Author as Shown One Year at the Detroit Exposition.

then the cases were turned so the long way of the cases faced out- wards, a fewer number of cases making a row that was slightly smaller than the others. Perhaps four rows were put up in this style, then they were again changed so that the narrow ends were outwards, which again reduced the size of the circle.   In this man- ner the size of the circles was gradually diminished as the pyramid increased in height, until its top was only two feet across.   That these cases might not be jarred out of place they were fastened to one another by means of small wire nails.   Upon the top of this pyramid was set a large number of two-pound, square bottles of honey. On top of the bottles was laid a platform of glass made by putting together two sheets of double-strength glass, bound together at the edges with cloth pasted on and covered with gilt paper. Upon the glass platform was set more bottles, then another sheet of glass a little smaller than the first one, and so on up, until a pyramid of extracted honey was constructed upon the top of the pyramid of comb honey, the former being surmounted by a huge boquet of golden rod. I remember building one such pyramid that was 16

Exhibit of Mr. M. H. Hunt, as Shown one Year at the Detroit Exposition.

feet in height. The spaces between the outer ends of the cases in the comb honey part of the pyramid was filled with small, "dime" bottles of honey. By thus combining the comb and extracted honey display, one "sets off" the other; in fact, my competitors sometimes complained of this, but it was their privilege to have taken advantage of this fact had they so chosen. Mr. M. H. Hunt one year had a castle in which the pillars were cases of comb honey piled up, and the balustrade was formed from panels of beautifully molded beeswax.

There is seldom a fairground with no bees near it, hence, no honey should be exposed. All honey should be shut up close, and no stickiness left on the outside of the package. Wax should be molded into fanciful shapes—statues, or something of that sort, if the exhibitor has the skill to make them. Fruits, vegetables, ears of corn, and the like, may be made of wax by first making molds, of plaster of Paris, from the objects themselves. It is not necessary that the articles be solid wax. First soak the molds in water, then pour in a small quantity of melted wax, close the molds, and then immediately shake them vigorously while the wax is cooling, thus coating the inside of the molds with wax. When the wax is cool it will come out all in one piece.

Let the beginner not try to show a multitude of things, but let what he does show be as good as it is possible for him to make it. Competition is so very keen, at least where the premiums are liberal, that it is folly to expect premiums upon second-class articles.

Now that I have reached the subject of premiums, it may be well to give what I would call a "model" premium list. I may have placed the premiums at higher figures than most societies would care to use, but the amounts can be easily reduced, preserving the proportions.

	1st	2nd	3rd
Most attractive display of comb honey	$35	$20	$10
Specimen of comb honey, not less than ten pounds, quality and manner of putting up for market to be considered	10	5	
Most attractive display of extracted honey	35	20	10
Specimen of comb honey, not less that ten pounds, quality and manner of putting up for market to be considered	10	5	
Most attractive display of beeswax	20	10	
Specimen of beeswax, not less than ten pounds, soft, bright yellow wax to be given the preference	6	3	
Single-comb nucleus Italian bees	10	5	
Single-comb nucleus black bees	10	5	
Single-comb nucleus Carniolan bees	10	5	
Single-comb nucleus Caucasian bees	10	5	

SWEEPSTAKES ON BEES.

	1st	2nd	
Display, in single-comb nuclei, of the greatest variety of the different races of bees	10	5	
Collection of queen bees of different varieties	16	8	

Honey vinegar, not less than one gallon, shown in glass	6	3
Assortment of honey candies......................	4	2
Pastry made with honey instead of sugar...........	4	2
The best specimens of honey producing plants, pressed and mounted, not to exceed 25 varieties........	15	8

SWEEPSTAKES.

The largest, best, most interesting, attractive and instructive exhibition in this department, all things considered...........................	35	20	10

I think bee-keepers would prefer to have "supplies" included in the premium list, but the managers of fairs have decided against the offering of premiums on such things, on account of the difficulty of securing satisfactory decisions. The best we can do is to have a "sweepstakes" premium offered upon the largest and best exhibit; *then* supplies will count.

A judge should never be compelled to take an exhibitor's word for anything. Let the article exhibited show for itself. Don't offer premiums on samples of different kinds of honey, when they can be so easily gotten up for the occasion by mixing. Don't put at the head of the list such requirements as: "Honey must be of this season's crop;" or "Must be the product of the exhibitor;" when there is no way of knowing whether they have been lived up to or not.

In my experience, one man to award the premiums, and he an expert, has given better satisfaction than three judges. It is difficult and expensive to get three men that are experts, and, even then the work is not always done so conscientiously, because it is not so easy to place the responsibility; each being able to shield himself behind the "other two."

Upon this point of judging, there is one other point often neglected that ought to be printed in connection with the premium list, and that is a "scale of points" for deciding in regard to the merits of exhibits. Particularly is this true in regard to honey. I would suggest the following: Color, 5; body, 5; flavor, 5; comb—straightness, 5; color of capping, 5; completeness of capping, 5; uniformity, 10; style, 10. Possible number of points, 50. By "uniformity" is meant the closeness of resemblance in the sections composing a specimen. "Style" including the attractiveness of the section and case; also the absence of propolis.

If a bee-keeper is going to make an exhibit of apiarian products, it often happens that he can also make exhibits in other departments of the fair, I have exhibited photographs in the art department, canned fruit in the fruit department, and the wife and children have sent things to their respective departments. In addition to this, when making a "circuit" of the fairs, I used to write them up for the Country Gentleman, getting paid for the work. Fairs come in the fall, after the busy season is over with the bees, and if a man has the time, taste and ability for this kind of work, going from one State to another, as I have done, he can probably clear $10 a day for five or six weeks in the fall. It is scarcely worth while to prepare for the work, however, unless there is some expectation of following it for several years.

Just a few parting words to the beginner: If you make an exhibit at a fair, don't get excited. Keep cool and have patience. Many unpleasant things *may* occur, but don't worry over them; and, above all, don't let the loss of expected premiums so "sour" you as to spoil your own enjoyment and that of your comrades. When you leave home have everything in readiness, as nearly as possible, to put right up. Pack everything carefully, but in such a manner that it can be quickly and easily unpacked. I used to pack the square bottles of honey in boxes furnished with partitions of cellular board, *a la* egg-crate, and, to pack the bottles it was only necessary to drop them into the openings, and nail down the cover. If the package does not indicate its contents, then mark it in some way. Never be compelled to open box after box in an exasperating hunt for something that *must* be had at once. And when fair is over, don't "go crazy" to get off the grounds the next minute. I have known of men sitting up all night swearing, and sweating and fuming, because "their car didn't come," or something of that sort, and we all went out on the same train the next morning. At the close of a large fair, an immense amount of goods are on the grounds; they have been several days accumulating, and it is *impossible* to move them all in an hour's time.

The Fertilization of Queens in Confinement.

EVERY experienced bee-keeper knows, of course, that the mating of a queen bee takes place upon the wing, in the open air; the queen leaving the hive for this purpose when about a week old. It seems to be Nature's plan for preventing in and in breeding; as the chances are that the queen will meet with a drone from some other colony than her own. There is no doubt that there is as much difference in our bees as there is in our other domestic animals, and the one thing that has prevented the development of strains of bees far superior to those we now possess, is that we have so little control over the choice of drones with which the queens shall mate. Where would our Shorthorns, Jerseys, Merinos, Poland Chinas and Plymouth Rocks have been now, if the mating of these animals had been no more under our control than is the mating of our queen bees ?

Attempts have been made to secure the desired object by confining young queens and drones in a tent, but, with one exception, some fundamental principle has been overlooked. For instance, drones of an improper age may have been employed; besides, a drone frightened to death and trying to escape from what, to him, is a prison, is in no mood to pay his addresses to a queen. The only wonder is that there has been an occasional success in carrying out this plan. The one really successful attempt at mating queens in a tent was made by Mr. J. S. Davitte, of Aragon, Georgia, and was described in the February issue of the Bee-Keepers' Review, for 1901. The mating tent was made as follows:

Twelve poles, about 30 feet in length, were firmly planted in the ground, 12 feet apart in a circle. From pole to pole, at the top, heavy wire was stretched to keep the poles true, and in place, and to afford support for the covering of mosquito-netting. Wires, or supports of some kind, are also stretched from each pole to its opposite neighbor. The poles are also braced from the inside. Common boards are used around the bottom to the height of five or six feet.

After the tent is complete, colonies of bees are placed up close against the wall of the tent, on the outside, each colony being given two entrances. One is the regular entrance, outside of the tent,

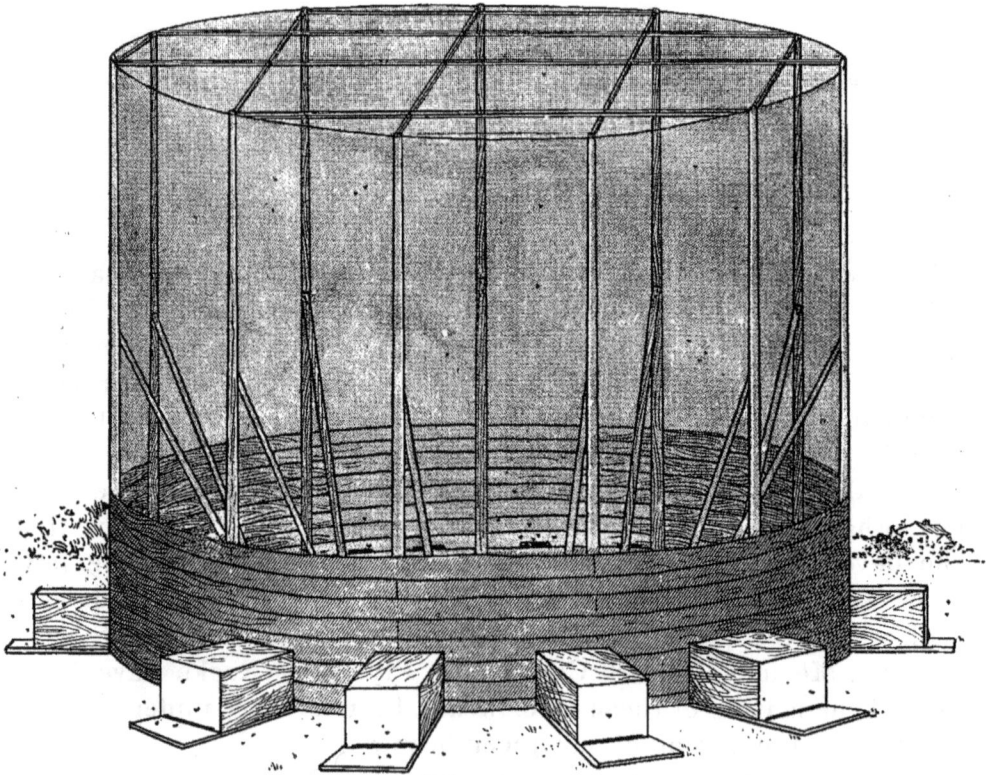

Tent for Controlling the Mating of Queens.

which is contracted by means of queen-excluding metal, so that neither drones nor queens can pass, but the workers can pass out and in, and work in the fields in the usual manner. The other entrance opens *into* the tent, and is large enough for the passage of a queen or a drone, but is kept closed or darkened for about a week after the colony is placed in position. This is done for the purpose of educating the workers to use the outside entrance. The drones are not allowed to use the outer entrance at any time, nor to enter the tent except from 11:00 a. m. to 1:30 p. m.   After the drones have learned the bounds of the tent they seem contented, and make a pretty school flying in the top of the tent.   Mr. Davitte says that the management of the drones is the main feature of the problem; once they become quiet, and reconciled to fly in the top of the tent, the problem is solved.   Nine times out of ten, the queen will not reach the top of the tent before receiving the most prompt and gushing attention.   The queens are not turned into the tent until the drones appear well-satisfied with the bounds of the tent, and, when in that condition, Mr. Davitte believes 500 queens in a day might be mated in such a tent.   One year he had about 100 queens mated in the tent. A queen would leave the mouth of the hive, and return in about five minutes, apparently mated, and, in three or four days, would be laying; and the progeny of all of the queens thus mated showed the same markings as the workers of the colonies from which the drones were taken.

When Mr. Davitte starts his queen cells, he places his colonies with selected drones around the tent, and allows them to fly in the tent a short time in the middle of each day, as has been explained, and, by the time the queens are old enough to be mated, the drones have become tamed, and so accustomed to their surroundings, or under control, so to speak, that, to quote from Mr. Davitte, "It would interest a bee-keeper to take his place inside the tent at noon, and see the ladies meet the gentlemen, who, Barkis-like are 'willin.'   I have seen the mating take place before the queen could reach the top of the tent.   Before they separate, the queen and drone fall nearly to the ground, and the queen goes directly to her home that she left not three minutes before."

As I look at the matter, the principal trouble with experiments in this line is that the drones have not been brought under control. When a drone has been accustomed to soar away in the blue ether for miles and miles, he is not going to be shut up in a 30-foot tent and be contented.   For a long time, at least, he is going to spend most of his time in trying to get out.   As I have already said, he is in no mood to pay his addresses to a queen.   Catch two wild birds at mating time, and shut them up in a cage.   Do you suppose that they would mate?   Canaries have been kept in captivity for many years. They are hatched and grow up in a cage.   They know no other freedom or life; and they mate in a cage.   Mr. Davitte had his drones flying for days in his tent before any queens were released in the tent.   Perhaps many of those drones had never flown in the outside air—knew nothing of it.   Having flown several days in the tent they became accustomed to that kind of flight, were in a normal condition, and ready to mate with a queen should one appear.

Suppose we could make a cage two miles wide, and half a mile high. Is there any doubt that a queen would be mated inside such a tent? Suppose it were reduced to one mile in width, and one-fourth of a mile high. Wouldn't it be a success? Let us go still farther, and have it half a mile wide and 80 rods in height. Isn't it reasonable to suppose that it would still be a success? The question then is: How small *can* it be and still be a success? My opinion is that the size is not so very material as it is to get the drones to fly and *feel at home*. One large enough for that is, in my opinion, large enough.

There is still one more point: Not all drones, at all ages, are capable of fertilizing a queen. Many experimenters have failed from not understanding this point. They have put nuclei, with young queens, into a tent, then caught drones at hap hazard and put them into the tent. Some of them may have been youngsters, just out of their cradles, so to speak. Others may have been old grey beards. All of them would certainly have been frightened out of their wits to have been thus caught and shut up in a tent away from their home. I think Mr. Davitte has found the key that will unlock the problem, viz., that of getting drones from a normal colony, that is working undisturbed in the open air, to fly *unworried* in an enclosure.

Although this account of Mr. Davitte's success was published some three or four years ago, I believe there has been nothing like it attempted since. The queen breeder who will build such a tent, and succeed with it as Mr. Davitte says that he succeeded, will certainly have one of the biggest advertisements that could possibly be secured. This is a matter that I should be glad to see some of the experiment stations take up.

Right in this line, it might be mentioned that Mr. L. A. Aspinwall, of Jackson, Michigan, reports success in clipping about ⅛ of an inch from the ends of the wings of a queen. While this does not prevent her flying, it so weakens her flight that she is not likely to go far from the apiary before being overtaken by an admirer. If only drones of a desirable character are allowed to fly in the home-yard, the chances are that the majority of queens will be well-mated. Mr. John M. Rankin, when at the Michigan Agricultural College, tried this same experiment, but, with him, only a small per cent. of the queens thus clipped ever became fertile. Perhaps he clipped them too much.

Here is an idea, however, that is thoroughly practical, one that can be put in practice by any bee-keeper: By the use of full sheets of comb foundation, or otherwise, keep the brood nests practically free from drone comb; then, in two or three, or half a dozen colonies (the number depending upon the size of the apiary) having very choice queens, allow an *abundance* of drone comb. The result will be that the air will be filled with drones from choice stock, and the majority of the queens will mate with these drones.

The Rendering of Beeswax.

IN nearly every apiary there are more or less odds and ends of combs which are well-worth saving to be made into wax. When an apiary is run for extracted honey the wax from the cappings is no small part of the income. If there are many combs to be rendered, as is often the case when foul brood gets into an apiary, the manner of doing the work becomes an important question. The small bee-keeper who has only a few scraps to melt up, may resort to almost any make-shift; and, by the way, here is one such primitive plan: Take an old dripping pan, or any large, flat, metal dish that is of little value, and punch a hole in one corner. Set the dish in an ordinary stove oven, letting the end with the hole in it project from the oven. Put the scraps of comb into the pan, where they will melt, and the wax will run out of the hole, where it may be caught in a dish set upon the floor. If the scraps are of nearly pure wax, like cappings, or new comb, this plan will answer quite well for rendering wax upon a small scale; but, if the combs are old, the cocoons will absorb so much of the wax that a large portion will thus be lost. A plan that will secure a larger percentage of wax from old combs, but requiring some more labor to put into operation, is that of crowding the combs into a sack made of cheese cloth or burlap, tying up the mouth of the sack, and immersing it in a boiler of water set upon a stove, and then bringing it to the boiling point. While the water and the sack and its contents are still hot, the sack should be thoroughly turned and pressed with something like a garden hoe, thus stirring up the contents and pressing out the wax. The water will largely take the place of the wax, which, being lighter than the water, will rise to the top, where it may be taken off in a solid cake after it has cooled. A weight of some kind, like a big stone, or some bricks, must be laid upon the sack to hold it at the bottom of the boiler while the wax is cooling, otherwise the sack will be embedded in the wax when it is cooled. This plan may be employed upon a large scale, even to the extent of using a large kettle out of doors, and the use of the sacks may be dispensed with by making a sort of pail or basket out of fine wire cloth, setting it down in the melted wax, inside the kettle, and then dipping off the wax with a dipper, by dipping inside the wire cloth basket, the wire cloth straining out the coarser impurities. This method of rendering wax by the use of boiling water will probably get out as much of the wax as it is possible to secure without the use of pressure upon the residue, or "slum gum," as it is called. Old combs

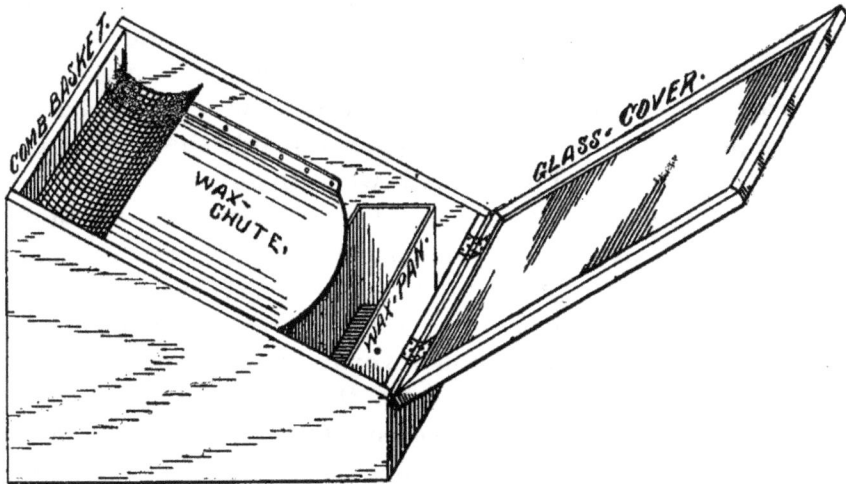

The Alpaugh, Solar Wax Extractor.

are largely made up of cocoons---more cocoons than wax---and they absorb and retain the melted wax, much as a sponge will hold water, and pressure is the only thing that will cause them to give up the golden treasure.

Another plan particularly feasible for melting cappings, new combs, or scraps that are nearly pure wax, is by the use of the solar wax extractor, which is simply a shallow box painted black inside and out, and furnished with a false bottom of sheet iron a few inches above the real bottom, a cover of glass completing the arrangement. The box is placed in a slanting position, facing the sun, and the refuse combs, etc., placed upon the false bottom of iron, or in a sort of basket arranged at the upper end for the reception of the cappings, scraps, etc. The direct rays of the sun, aided, sometimes, by the reflected rays from the cover, to which is fastened a sheet of bright tin, melt the wax, and it runs down to the lower end of the metal shute where it drops off into a vessel set there to catch it. A small solar wax extractor standing in an apiary is an excellent thing, as into it may be thrown all scraps of comb that would otherwise be thrown away, or perhaps be thrown into a box or barrel to stand around until destroyed by the bee moth's larvae.

All of these plans of rendering wax fall short of perfection, however, as too much wax is left in the residue. Pressure of some sort must be used, or a large part of the wax is lost. For making small, or ordinary quantities of wax, what is called the German wax press answers the purpose quite well. This is a tall can made of heavy sheet-metal, with a false bottom securely fastened to the sides a few inches above the real bottom of the can, together with a screw and follower above to bring pressure to bear upon the mass of combs after the wax has been thoroughly melted by the steam that arises from the water that has been placed below, previous to setting the can upon a stove. For holding the combs, a wire cloth basket is used and a piece of cheese cloth is placed inside the basket before putting in the combs. The melted wax drips down upon a false bottom and runs out through a spout that passes out through the side of the can. The use of pressure while the slum gum is still surrounded by live steam secures nearly all of the wax; and the greatest objection to the use of the German press is its limited capacity—it is too slow a process if much work is to be done.

Extracting the melted wax from the slum gum by means of centrifugal force, the same as syrup is thrown from the sugar in a sugar refinery, or water from clothes when dried in a laundry, has been tried enough to enable us to say that something may be hoped for in this direction.

For making large quantities of wax, probably the most practical plan is that of melting up the cappings or combs in a boiler or large kettle, dipping off the wax from the top, and putting the slum gum through a powerful press. It is difficult to say exactly who first utilized screw-power for pressing the wax out of slum gum, but I think C. A. Hatch, of Richland Center, Wisconsin, was the first to bring the matter prominently before the public. He was followed by Mr. F. A. Gemmill, of London, Ontario, Canada. One form of press is, I believe, now called the Hatch-Gemmill press. Of course, such presses may vary in detail, and I think the best form that I have ever seen was illustrated and described in the Bee-Keepers' Review by Mr. E. D. Townsend. Here is his description of the press and w he would use it.

The Hatch-Gemmill-Townsend Wax Press.

"Procure two pieces of tough oak 3 x 4 inches, by 24 inches long. (See Fig. 1.) Twenty inches from center to center, at equal distances from each end, bore ¾-inch holes through the 3-inch way of both pieces. These holes are for the ¾ rods to pass through to form the main uprights. Then in the center of one wooden piece, parallel with the other holes, bore a 1⅛-inch hole. This is for the bench-screw to work through. From a ¾-inch iron rod, have a blacksmith cut two pieces 20 inches long, and one 30 inches long. The two, 20-inch pieces are to have threads cut for a distance of 5½ inches on both ends; and each piece is to be furnished with four burrs and four washers. To assemble the machine, turn a burr clear on at each end of each rod. Next put on a washer, on each end, and slip the ends of the rods through the holes in the ends of the wooden pieces; then put a washer on over the projecting end of each rod, over this a burr, and adjust the parallel, 3 x 4 pieces a scant 13 inches apart.

When the ⅞-inch bottom, or table, is on, there will be 12 inches space, in the clear, between the top of the press-table, and the under side of No. 1. The screw is the same as a carpenter uses in his work-bench vise. Mine is 16 inches in length, and 1 1-16 in diameter. The burr for the screw to work in is let into the under side of No. 1, and held in place by a 3 x 12-inch steel plate, ¼-inch thick, with a 1 1-16-inch hole in the center for the screw to work through. Reference to the accompanying engraving will show the bolts that hold this plate in position. Instead of having round holes in the ends of the plate, for the bolts to pass through, they are made in the form of slots that extend crosswise of the plates. There should also be another plate on top, only the slots in the ends ought to extend the other way—lengthwise. This arrangement allows of any adjustment of the screw so that it will stand perpendicularly. The strain here is something enormous, and everything must be made solid.

The frame can be made to suit one's fancy; mine is of 2 x 4, well-braced, 24 inches high, the top 24 x 26 inches, the long way parallel with the 3 x 4 piece. Don't forget to brace it well with a ⅜ rod from the table to No. 1. (See cut.)

The 30-inch, ¾-inch rod is for a lever for turning the screw. The lever that comes with the screw is not sufficient.

No. 5 is a pan of galvanized iron, five inches deep, and 18 inches square, with one side left open and formed into a spout to carry off the wax.

There are two racks (No. 2), each 16 inches square, made of one-inch, square pieces of pine, spaced ⅜ of an inch apart, and cleated at each end.

The follower (No. 3) is of the same size, and is made of two thicknesses of ⅞-inch boards, with the grain running in opposite directions. In the center of the follower, on top, is a 3 x 5-inch steel plate, with an indentation in the top for the screw to work in. Two screws fasten this plate to the follower.

The form (No. 4) is 15 inches square, and five inches deep.

In the rendering of wax, there is an excellent reason, which will be given later, for rendering the cappings separately from the old combs. At present I will describe the work of rendering the cappings separately from the old combs. Put a pail of clean, soft water into a No. 9 wash boiler, and set it over a slow fire. Fill the boiler full of cappings; and, as they melt down, add more, until the boiler is as full as it can be handled conveniently—say, within two inches of the top. If the cappings are broken up fine they will melt much quicker. My cappings are all stored in cracker or sugar barrels. I throw a barrel into a box that will hold five or six bushels, take an old axe and cut off the hoops and remove the staves, then chop up the cappings with a spade.

We will suppose that you have a boiler of wax on the stove. See to it that the fire is not too hot. If you *have* a hot fire, leave the griddles on the stove under the boiler. Keep the lumps broken up with a long paddle—a barrel stave will answer—and keep constantly in mind that the wax should *never be allowed to boil.* Just as soon as the last chunk is melted, slide the boiler off the stove, upon a barrel or box arranged the same height as the stove. The wax is now ready for the press. See that the press-screw is clear up out of the way, and the galvanized iron pan (with the spout end) is in place. Next put in one of the slated frames, then the form, over which spread a 30-inch, square piece of burlap of the thin, open kind. Press the burlap down into the form, set a galvanized iron washtub under the spout (you will need three or four of these tubs) then, with a large dipper, having a long handle, dip the wax from the boiler to the press. By the way, it is not necessary to put all of the melted wax through the press. With a little care all of the slum gum can be dipped off the top, leaving quite a quantity of wax, water and honey that can be emptied directly into the tub. When the form has been filled, take hold of the two opposite sides of the burlap, and move it up and down; then do the same with the two other sides. This works most of the water and wax through the burlap, out of the way, so that we can handle the slum gum to better advantage.

We will suppose that nearly all of the wax is out that will come out without pressure; take hold of the burlap on two opposite sides, bring them together with a good lap, and pin with a ten penny nail;

then handle the two remaining sides in the same manner. Next remove the form, and put another rack on the top, the same as was used underneath. Now put on the follower and add the screw-pressure. When the wax stops running, loosen up the screw, give the cheese a half turn, and add more pressure. If you have done a good job there will not be a *particle* of wax left in the slum gum. One pressure is all that is necessary for a boiler full of cappings; but with old combs four or five times may be required. Always keep in mind that the less slum gum put into the press the more perfectly can it be freed from wax.

It will be noticed that the honey and wax have never been brought to the boiling point, hence the honey has not been injured for the making of vinegar; and after the wax has cooled and been taken off in a cake, the honey and water may be emptied into an open-end barrel. After it has stood over night, or until the sediment has settled, skim, and dip off the top, and the sweetened water thus secured is as good material for making vinegar as it is possible to secure; while every ounce of wax has been removed from the cappings. The sweetened water thus secured is *too* sweet for the making of vinegar, but more soft water may be added and the vinegar made in the usual way. This is why we do not render old, black, brood combs at the same time that we melt up the cappings.

The same boiler is used for clarifying the wax. After a little cleaning around the upper edge, put in a pail of water, then fill with the wax as it comes from the press, only be particular to chop it up fine. Any chunks larger than two inches in diameter should be chopped up with the axe, as we wish to melt it with the *least possible heat*. As in the first melting, the chunks are kept broken apart with a paddle. Give a little more time for the wax to melt rather than have it boil; and just the moment that it is all melted, slide it off the stove the same as before, cover up with two or three thicknesses of blankets, and let it stand until there are signs of its caking around the edges. Usually, four or five hours are enough time for the impurities to settle to the bottom. After skimming the wax, it is ready to dip off and cake. Clean your long handled dipper, and with it dip off the wax into flaring-top, tin pails. Don't make the mistake of putting any water into the pails. There is a little knack about dipping out the wax in such a way as not to disturb the sediment any more than is possible. Don't think of dipping right in, just as though you were dipping water, but drop the side of the dipper into the wax, say, three-fourths of an inch, then carefully sink the *bottom* of the dipper down into the wax, always keeping the *top* edge near the surface of the wax. By dipping in this manner, it is surprising to see how close one can dip to the sediment without disturbing it. Stop dipping as soon as signs of sediment appear in the dipper, and what is left in the boiler can go into the next melting. Let me repeat: If you wish for nice, soft, pliable wax, that is so much in demand in the markets, *don't ever allow your wax to boil* in any process of rendering."

For cleaning any utensils that are daubed up with wax, use a cloth saturated with benzine. Benzine will dissolve wax much as water will dissolve sugar.

For some mysterious reason, sulphuric acid will cleanse or clarify beeswax that is brown, or black, or almost any color, bringing it back to a nice, bright yellow. The bee-keeper who renders his wax according to the methods here described, will probably have no need for using acids, but those who buy wax for making into foundation find the use of the acid almost indispensible. A kettle or some other vessel, is filled perhaps one-third full of water, and then filled up with cakes of wax. By the use of steam, or by setting the vessel on a stove, the wax is melted, when acid is added at the rate of about one pint of acid to 12 gallons of water. Soon after the acid is poured in, the wax will be seen to change to a lighter hue, when the heat may be stopped, and the sediment allowed to settle, after which, the pure wax can be dipped of the top. If a metal vessel is used, it must be thoroughly washed after use, and it would be well to rub it over with grease to prevent any further action of the acid.

The Relation of Food to the Wintering of Bees.

IN the Southern States, and other places not blessed with a stern winter, where bees can enjoy frequent flights, it matters little what the food is, so long as it is not actually poisonous. By this is meant that any kind of sweet like sugar, honey, or even honey dew, will answer as food. In these mild climates, little or no protection is needed; but, as higher latitudes are reached, chaff-packed hives, or their equivalent, are needed, and there must be some care exercised in regard to food. As we journey still further from the equator, it is only cellars and the best of food that bring forth uniform results.

It has been asserted that honey is the "natural" food of bees, and that nothing can be gained by substituting something else. It must be remembered that the "natural" home of the bee is that of a warm climate, where there are no long spells of confinement caused by continued cold. Honey is, of course, the "natural" food of bees, but this fact does not prevent their dying sometimes as the result of its consumption, when a diet of cane sugar would have saved their lives.

In my opinion, food is the pivotal point upon which turns the wintering of bees in our Northern States. Food is the fulcrum, and temperature the long end of the lever. The whole question in a nut shell is just this: The loss of bees in winter, aside from that caused by diarrhea, is not worth mentioning. It is *diarrhea* that kills our bees. What causes it? An overloading of the intestines, with no opportunity of emptying them. Cold confines the bees to their hives. The greater the cold the larger are the quantities of food consumed to keep up the animal heat.   The more food there is consumed, the sooner are the intestines overloaded.   A moment's reflection will make it clear that the character of the food consumed has an effect upon the accumulation in the intestines.   In the digestion of cane sugar there is scarcely any residue.   Honey is usually quite free from nitrogenous matter, being well supplied with oxygen, and, when practically free from floating grains of pollen, is generally a very good and safe winter-food; although not as good as properly prepared sugar syrup, which never contains nitrogen, but possesses more oxygen.   The excreta from diarrhetic bees is almost wholly pollen grains, in a digested or partly digested state, with a slight mixture of organic matter.   What overloads the intestines of the bees is this nitrogenous matter which they consume, either as grains of pollen floating in the honey, or by eating the bee bread itself.

Repeated experiments have proved beyond a doubt that, as a winter food for bees, cane sugar has no superior.   With this as an exclusive diet, bees never die with the dysentery; and, if kept in a temperature ranging from 35 to 42 degrees, they are all but certain to winter successfully.   This being the case, the question naturally follows, why not take away the honey in the fall, and feed the bees sugar?   One objection to the use of sugar, as a winter food, is that every pound of sugar so used puts one more pound of honey on the market.   Another objection is that the bee-keeper is thereby compelled to pay out money for sugar, while he may have on hand a crop of honey that is meeting with slow sale.   Some object to its use on the ground that it lends color to the cry of "adulteration."   Perhaps the greatest objection is the labor of extracting the honey and feeding the sugar.

Let's consider these objections.   The use of sugar as a winter food for bees unquestionably does put a little more honey on the market, but this ought not to weigh so very heavily against the certainty of wintering the bees.   Neither need there be any labor of extracting the honey in the fall, if the summer management has been conducted with a view to feeding sugar in the fall.   By contraction of the brood nest nearly all of the honey may be forced into the supers, leaving the brood combs nearly empty at the end of the season.   It only remains to feed the bees, and, with proper feeders (the Heddon, for instance), tin cans, and oil stoves for making the syrup, feeding is neither a long nor a tedious task.   What little honey remains in the corners of the combs is not likely to be consumed until spring, when frequent flights will prevent all troubles that might arise from its consumption.   In regard to causing the public to believe that by some hocus pocus the sugar that is fed gets into the surplus, no one need know of the feeding, except it might be in some cases, an immediate neighbor, and the bee-keeper ought to enjoy his neighbor's confidence to that degree that the exact truth can be told him, and it will be believed. As in regard to the increased amount of honey that the use of sugar as winter stores puts upon the market, so any possible talk about adulteration is over-balanced by the certainty of carrying the bees through the winter.

If the feeding is done early enough so that the bees will have time to work the honey over and ripen it, no heat will be needed in making the syrup; simply stir into cold water all of the sugar that it will dissolve, feed it to the bees, and they will reduce it to the proper consistency; and, by the addition of their secretions, change the cane sugar into grape sugar, thus practically making it into honey. If fed too late something may be necessary to prevent the granulation of the syrup. For this purpose I never found anything better than honey—from 10 to 20 per cent. is sufficient. September is early enough to feed; but, when feeding *has* been neglected until it is so late and the weather so cool that the bees will not leave the cluster and go into the feeder, it may be managed, as explained in the chapter on feeding, by filling the feeder with hot syrup and placing it *under* the hive. The heat from the syrup will warm up and arouse the bees, when they will come down and carry up the feed.

But all can not, or may not wish to, use sugar for winter stores, and many do not *need* to use sugar to insure the successful wintering of their bees. There is a great difference in localities regarding the character of the honey. Where one has successfully pursued the same course year after year, it is doubtful if a change would be desirable; but what shall the man do who loses heavily nearly every winter, yet can not, or will not use sugar? Possibly he can so manage that his winter stores are secured from a different source. Mr. O. O. Poppleton takes the ground that the best winter stores come from the most bountiful yields. It is possible that there is something in this, bountiful yields of any crop are usually of fine quality, but I know of at least one locality where the fall flow of honey is always the most abundant, and I might almost say *always* abundant, yet so surely will it kill bees that the most extensive bee-keeper in that locality, after an experience of many years, kills his bees in the fall rather than attempt to winter them on this honey by *any* method.

But bee-keepers can do this: Notice if any particular kind
of honey is more likely to cause trouble, and then avoid its use as
winter stores. Part of the bees may be protected upon the summer
stands, and part put into the cellar. In a warm, open winter, the
bees out of doors will stand the better chance; in a severe winter the
odds will be in favor of the cellar—and their owner must take his
chances.

Out-Door Wintering of Bees.

IF bees can enjoy frequent flights, out of doors is the place to winter them. If deprived of these flights, a temperature of about 45 degrees enables them to bear a much longer confinement than does a temperature below freezing. In the South, frequent flights are assured; in the North, no dependence can be placed upon the matter. Some winters are "open," or there are January thaws, allowing the bees to enjoy cleansing flights, while other winters hold them close prisoners for four or five months. It is this element of uncertainty attending the wintering of bees in the open air that has driven so many bee-keepers to the adoption of cellar wintering. Still, there are some bee-keepers who, from some peculiarity of location or management, are able to winter their bees in the open air with quite uniform success; others are *compelled*, for the present, at least, to winter the bees out of doors; in short, a large portion of the bees, even in the North, are wintered in the open air, and probably will be for a long time to come; and, while my preference is the cellar, there is much to be said in favor of out-door wintering. Let me give one or two instances of success: Ira D. Bartlett, of East Jordan, Michigan, which is away in the northern portion of the lower peninsula, began keeping bees when only 14 years of age—began with only one colony—and when 21 years of age he had 150 colonies, and had never lost a colony wintering them out of doors. His method of protection was very thorough. He packed four colonies in one box, putting packing not only at the sides, and on top, but also *below* the hives. The packing was fine, dry sawdust, and the roof to the box had eaves that extended over like the eaves of a railroad station, which allowed the roof to be raised up a

The Old Home of Two Lady, Pioneer Bee-Keepers.

short distance above the box, for ventilation; yet the snow would not get in to any great extent. There was a sort of vestibule in front of the entrances, and this vestibule was kept closed by means of a board; it being removed only when there came a day warm enough for the bees to fly—something that rarely occurred in the winter. So warm and comfortable were the bees when so snugly housed that they even brought the dead bees out and dropped them in the vestibule. I suspect that the thorough protection, combined with the perfect ventilation, allowing no accumulation of moisture, is the secret of this wonderful success.

Another instance was that of two ladies who began bee-keeping in Northern Michigan before the iron horse had invaded that region, and who were uniformly successful, for a long series of years, in wintering their bees out of doors, packed in chaff. Like Mr. Bartlett, they furnished abundant upward ventilation, above the packing. An opening a foot square was cut in the top of the box containing chaff that was placed over the colony, and this opening was covered with wire cloth to keep out mice; and then, over all, to keep out the storms, was a large roof. So successful were these ladies, that, from first to last, I have paid them nearly $1,000 for bees.

It does not seem as though the question of whether bees should be protected, here in the North, need receive any consideration whatever, yet it has been objected to on the grounds that the packing becomes damp, that it deprives the bees of the warmth of the sun, and that they sometimes fail to fly in the winter, because the outside warmth is so slow in reaching them, when bees in single-wall hives may be in full flight. There is occasionally a still, mild day in winter, upon which the sun shines out bright and strong for an hour or two, and bees in single-wall hives enjoy a real cleansing flight, while the momentary rise in the temperature passes away ere it has penetrated the thick walls of a chaff hive. On the other hand, there are days and weeks, and sometimes months, unbroken by these rises in temperature; and the bees must depend for their existence upon the heat generated by themselves; and the more perfect the non-conductor by which they are surrounded, the less will be the loss of heat. When bees are well protected, there is less necessity for flight than when the protection is slight. If a bee-keeper thinks, however, that bees in a chaff hive ought to fly on a warm day, and they *don't* fly, he has only to remove the covering over the bees, and allow them to fly from the top of the hive.

For several winters I left a few colonies unprotected; and I discontinued the practice only when thoroughly convinced that, in this locality, the losses were lessened by protection. In mild winters the bees came through in pretty fair condition. In severe winters the bees in the outside spaces, or ranges of combs, died first; the cluster became smaller; the bees in more ranges died; and, by spring, all were dead, or the colony so reduced in numbers, and the survivors so lacking in vitality, as to be practically worthless.

I have never seen any ill effects from dampness, but I have always given abundant ventilation above the packing. When the warm air from the cluster passes up through the packing, and is met by the cold, outer air, some condensation of moisture takes place. This moistens the surface of the packing, but it remains comparatively dry underneath. With a good strong colony of bees, and ventilation above the packing, I have never known trouble from moisture.

In the giving of protection, chaff hives have the advantage of always being ready for winter, and of doing away with the labor and untidiness of packing and unpacking; but they are expensive and cumbersome. It is some work to pack bees in the fall, and to unpack them again in the spring, but light, single-wall, readily movable hives during the working season are managed with enough less labor to more than compensate for that of packing and unpacking. Then there is another point: The work of packing and unpacking comes when there is comparative leisure, while the extra work caused by great, unwieldy hives, comes at a time when the bee-keeper is working on the keen jump.

For packing material I have used wheat chaff, forest leaves, planer shavings and dry sawdust. I have never used cork-dust, but it is probably the best packing material. Its non-conductivity is nearly twice that of chaff, while it never becomes damp. The only objection is that it is not readily obtainable, and usually costs something, while the other substances mentioned cost nothing. What they lack in non-conductivity can be made up in quantity. And this brings up the point of the proper thickness of packing. I have often thrust my hand into the packing surrounding a populous colony of bees, and found the warmth perceptible at a distance of four inches from the side, and six inches from the top. This would seem to indicate the thickness when sawdust or chaff is used. I presume that packing has been condemned when it was not more than half done—that is, when not enough material is used. I don't appreciate the arguments of those who advocate the use of *thin* packing. I don't believe that the benefit of the heat from the sun during an occasional bright day, can compensate for the lack of protection during *months* of extreme cold.

Packing Bees for Winter in Long Boxes.

Hollow walls with no packing have had their advocates; and it has been asked if these dead (?) air spaces were not equally as good non-conductors of heat as those filled with chaff. They are not. In the first place, the air is not "dead;" it is constantly moving. The air next the inside wall becomes warm and rises; that next the outer wall cools and settles; thus there is a constant circulation that robs the inner wall of its heat.

If chaff hives are not used, how shall the packing be kept in place? I know of nothing better than boxes made made of cheap lumber. If there is lack of room for storing them in summer, they can be made so as to be easily "knocked down," and stacked up when not in use. Of course, bees can be packed more cheaply by setting the hives in long rows, building a long box about them, and filling it with the material used for packing. With this method the packing ought to be postponed until it is so late that the bees are not likely to fly again until they have forgotten their old locations; else some of the bees will be lost, or some of the colonies get more than their share of bees. When they have a "cleansing flight" in winter, there is also a likelihood of some bees returning to the wrong hive. Then, when the bees are unpacked in the spring, there is more confusion and mixing; but I don't look upon this as so very serious a matter. At this time of the year, other things being equal, a bee is worth just as much in one hive as in another. If there is any difference in the strength of colonies, the weaker ones might be left nearest to where the bees were unpacked.

Speaking of being compelled to wait about packing the bees until they are not likely to fly again until some time in the winter, reminds me that advantages have been claimed for *early* packing; that bees in single-wall hives only wear themselves out with frequent flights that are to no purpose, while those that are packed are not called out by every passing ray of sunshine; that the early-packed bees sooner get themselves settled down for their winter's nap, and are in better condition for winter when it comes. It is possible that there is something in this, but there were two or three years in which I tried packing a colony or two as early as the first of September, and I continued to pack a colony every two or three days until the fore part of November, and I was unable to discern any advantage in very early packing. If the bees are protected before freezing weather comes, I believe that is enough.

There is one other point that ought not to be neglected in preparing the bees for winter, whether in-doors or out, and that is the leaving of a space below the combs. When wintered out of doors there ought to be a rim two inches high placed under each hive.

This not only allows the dead bees to drop away from the combs to a place where they will dry up instead of moulding between the combs, but if there is an entrance cut in the *upper* edge of the rim, there will be no possibility of its becoming clogged. This empty space under the combs seems to have a wonderful influence in bringing the bees through in fine condition, and I am not certain *why*.

Weak colonies can seldom be wintered successfully out of doors. They cannot generate sufficient heat. In the cellar, where the temperature seldom goes below 40 degrees, quite weak colonies can be successfully wintered.

As I understand it, this whole matter of out-dooor wintering of bees might be summed up in a few words: Populous colonies; plenty of *good* food and *thorough* protection. Simple, isn't it? Yet there is a world of meaning wrapped up in those few words.

The Ventilation of Bee Cellars.

YEARS ago, "sub-earth" ventilation of bee cellars was almost universally recommended. Nearly every one who built a bee cellar, also buried 200 or 300 feet of drain-tile; the outer end connecting with the open air, and the inner end entering the cellar. To remove the air *from* the cellar, a pipe, connecting with a stove pipe in the room above, extended down through the floor to within a few inches of the cellar bottom. The draft of the stove pipe "pulled up" the air from the cellar, and more flowed in through the sub-earth pipe to take its place. In passing through the sub-earth pipe, the air was warmed. If there was no stove pipe with which to connect the outlet pipe, it was extended upwards until it reached the open air. The air in the cellar, being warmer than the outside air, flowed out of the upper ventilator, and more air flowed in through the sub-earth tube.

In order to keep the temperature even, there was much opening and closing of the ventilating tubes. In very severe weather, it was often necessary to leave the openings closed several days, or even weeks. At such times it was noticed that the bees suffered no inconvenience. Not only this, but it was often noticed that when the ventilators were opened, the in-rush of fresh, cool air aroused the bees and made them uneasy. Finally, the ventilators were opened less and less, and, at last, they were left closed all of the time.

The amount of air needed by bees varies greatly according to circumstances. When they are excited and full of honey, as is the case with a swarm, the amount of air needed is very great. If they can be kept quiet, a very little air will suffice. In winter the bees are in a semi-dormant state, one closely bordering on hibernation, as that word is properly understood, and the amount of air necessary for their maintenance is very slight. I believe it was Mr. D. L. Adair who, years ago, removed a box of surplus honey from a hive, and, leaving the bees in possession, pasted several layers of paper over the entrance. As all of the cracks and crevices were stopped with propolis, the box was practically air-tight. The bees were kept confined several days, yet did not, apparently, suffer for want of air. Mr. James Heddon tells of some man who, wishing to "take-up" some of his colonies in the fall, plastered up the entrances with blue clay, expecting to kill the bees by suffocation. Upon opening the hives a few days later, imagine the discomfiture of their owner at seeing the bees fly right merrily. I have several times wintered bees successfully in "clamps" where the bees were buried under two feet of frozen earth. Prof. A. J. Cook even went so far as to hermetically seal up two colonies by throwing water over the hives and allowing it to freeze, thus forming a coating of ice over the hives. The bees survived this treatment. It is not likely that, in any of these experiments, the coverings surrounding the bees were absolutely air tight, but enough is proven to show that, in winter, bees can survive, and, apparently thrive, with a very limited amount of air.

Special ventilation, simply for the sake of securing fresher or purer air, seems to be almost wholly unnecessary; the few bee-keepers who plead for special ventilation do so almost wholly upon the ground that they can thereby more readily control the temperature. If the repository is sufficiently under the ground, it does not seem as though ventilation would be very much needed for controlling the temperature, unless it might be towards spring when the bees had commenced breeding, and a large number of colonies were in the cellar.

When bees settle down into that quiescent state that accompanies successful wintering, their need of air is very slight, indeed. When their winter nap is ended, and spring arouses them to activity, and to brood rearing, more air is needed. It is then, if ever, that special ventilation is a benefit, but, as this can be secured, in the ordinary cellar, by the opening of doors and windows at night, if it ever becomes really necessary, it scarcely seems necessary to go to the expense of supplying sub-earth pipes. I should not do it, nor advise it. When bees are to be wintered in large numbers, in a special repository, I would have some arrangement whereby the heat could be allowed to pass off, if it should become advisable, yet not allow the entrance of light.

The Relation of Moisture to the Wintering of Bees.

IS IT an advantage to have the air of our bee cellars dry? Or, do the bees winter more perfectly in a moist atmosphere? Or, is this an unimportant factor? If it is important, how shall we determine what degree of moisture is most conducive to the health of the bees, and, having decided this point, what shall we do about it? How can we control the amount of moisture in the air of our bee cellars? All these queries, and many more, come to the man who is thinking of wintering his bees in a cellar.

Whether bees can be successfully wintered in a damp cellar, depends largely, almost wholly, upon the *temperature* of the atmosphere. "If the repository be damp, a degree of temperature higher in proportion to the dampness should be maintained."—*N. W. McLain*. Referring to this statement, Mr. Frank Cheshire says: "The reason being that the water has an enormous capacity for heat (specific heat) whether in the liquid or vaporous form; the latter abstracts heat from the bees, and intensifies their struggle." Dr. Youmans says "Air which is already saturated with moisture refuses to receive the perspiration offered it from the skin and lungs, and the sewage of the system is dammed up."

A moist air very readily absorbs heat, and more quickly robs the bees of that element so essential to life; hence it will be seen why a moist atmosphere must also be a warm one if disastrous results are to be avoided.

There is also another point, in the wintering of bees, upon which moisture has a bearing, and that is in regard to its effects upon the exhalations of the bees. If the exhalations are not taken up readily, the "sewage of the system is dammed up." But little moisture is required to saturate cold air; that is, it will absorb but little moisture, the point where it will receive no more being soon reached. As the temperature rises, the absorbing capacity of the air increases. When air of a high temperature, at that of our bodies, or nearly that, is saturated, or nearly so, with moisture, the exhalations from the lungs and skin are taken up but slowly; we are oppressed, and say the weather is "muggy." This explains why we feel better on bright, clear days. Heating air increases its power of absorption, hence we enjoy a fire on a damp day. If the air of a cellar is dry, it will be readily seen that the temperature may be allowed to go much lower. In other words, a cold, dry atmosphere or a damp, warm one, may be about equal, so far as effects are concerned. This is a point that bee-keepers have not sufficiently considered.

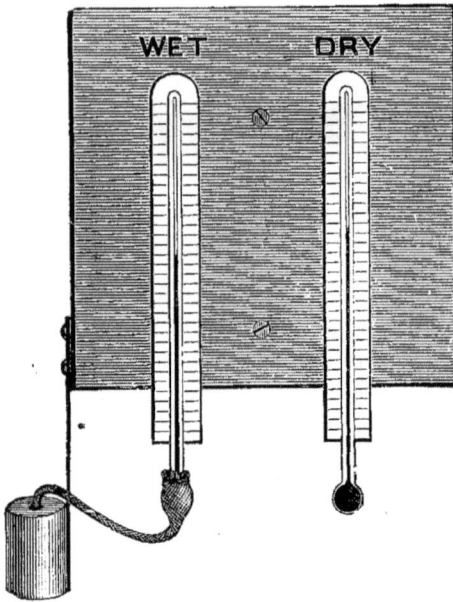

We have many reports of the successful wintering of bees at such a degree of temperature, but nothing is ever said as to the degree of *saturation*. Bee-keepers ought to use a wet-bulb thermometer in their cellars; then let the degree of saturation be given with that of the temperature, and we would have something approaching accuracy. I say "approaching accuracy," because the strength of the colonies, and the manner in which they are protected, have a bearing. A populous, well-protected colony can warm up the inside of the hive, expelling the moisture, and increasing the absorbing capacity of the air. Building a fire in a room on a damp day is the same thing in principle.

As mentioned in the preceding paragraph, the way to decide in regard to the amount of moisture in the air, is by the use of a wet-bulb thermometer. The arrangement is very simple, and any of my readers could make one. Attach two ordinary thermometers, side by side, to a piece of board. Just below them, fasten a tin cup for holding water. Make a light covering of candle wicking for one of the bulbs at the bottom of the thermometer, allowing the wicking to extend down into the water in the cup. The water will ascend the wicking and keep the bulb constantly wet. There will be, of course, evaporation from the wick surrounding the bulb. Evaporation causes a loss of heat; hence, the drier the air the greater the evaporation, the greater the loss of heat, and the lower will go the mercury in the wet-bulb thermometer. The greater the difference in the readings of the wet and the dry bulb thermometers, the drier the air. In the open air there is sometimes a difference of 26 degrees; but this is unusual. When it is raining, the air is then saturated. There is then no evaporation, and both thermometers show the same degree of temperature. In the cellar in which I used the wet-bulb thermometer the difference in the readings of the two thermometers was usually about three or four degrees, with the wet-bulb instrument standing at about 36 degrees; but this difference could be increased two or three degrees by warming the air with an oil stove. If the mercury in the wet-bulb thermometer stands at 36 or 40 degrees, and that in the dry-bulb as much as four degrees higher, I think there need be no worry about moisture; but if the difference is only two degrees or less, either the temperature ought to be raised, or the air dried in some manner.

Ventilation of cellars has been objected to on the ground that it brought moisture into the cellar. This may be true, but not in freezing weather. Frozen air, if the expression is allowable, has a very low point of saturation. That is, it will hold very little moisture; and when it is brought into the highe1 temperature of the cellar, and becomes warmed, its capacity for absorption is greatly increased— it is ready to receive water instead of giving it out. When the outside air comes into the cellar, and deposits moisture upon objects therein, it is evident that the in-coming air is warm and moisture-laden—warmer than the cellar and its contents.

Mould in bee-repositories is usually looked upon as something undesirable, and I will admit that its appearance is far from pleasant, but we must not forget that, in a certain sense, it is a plant— the child of warmth and moisture—and that the conditions necessary for its development may not be injurious to the bees—*may* be more beneficial than a condition under which mould does not develop, viz., one of moisture and *cold*. A very damp cellar ought to be warm enough for the development of mould. But the cellar need not be damp. It can be made both warm and dry. These matters of temperature and moisture are under our control. Either by fires, or by going into the earth, preferably the latter, we can secure the proper temperature; and by the use of lime to absorb the moisture, a dry atmosphere can be secured. Certainly, it is not much trouble to keep unslacked lime in the cellar. A bushel of lime absorbs 28 pounds of water in the process of slacking.

While it is evident that moisture in ordinary cellars is not injurious, provided the temperature is high enough, it is a great comfort to know that there is nothing to fear from a dry atmosphere; that we can indulge our fancy, if you choose to call it that, for dry, sweet-smelling, mouldless cellars, and know that the results will be harmless.

Some bee-keepers have asserted that cellars dug in clay or hard pan are more difficult to keep dry than when dug in a sandy soil. Mr. J. H. Martin, when living in New York, said that a cellar in hard pan, or even in clay, could be much improved by digging down two or three feet, filling in with stones, then with gravel, and finishing up with a covering of cement.

A Glimpse of a Montana Apiary.

The Influence of Temperature in Wintering Bees.

PROF. ATWATER says that the production of heat in the human body is so great that, if there were no way for it to escape, there would sufficient accumulate, in an average, well-fed man, to heat his body to the boiling point in 36 hours. This heat is gradually passing off by radiation. To prevent too rapid radiation, we cover our bodies with clothing. For the same reason, we surround our bees in winter with chaff or some other non-conductor of heat; but there is no way in which the radiation of heat can be so completely controlled as by surrounding the heat producing body with an *atmosphere* of the proper temperature. There is no method by which the most desirable temperature for wintering bees can be so completely secured as by placing the bees in a cellar or special repository.

The ordinary house-cellar, where the temperature remains above freezing, is usually a good place in which to winter bees. Men who are engaged extensively in bee-keeping where cellars are needed for the wintering of bees, usually find it to their advantage, perhaps a necessity, to build a special repository. The more completely the cellar is below the surface of the earth, the more perfectly can the temperature be controlled. It should be remembered that, not only is there the winter's cold with which to contend, but the warmer days of late winter may arouse the bees and make them uneasy before it is time to remove them from the cellar; unless the cellar is deep in the ground beyond the influence of outside temperatures. The walls of a cellar are usually laid up with brick or stone, but there are other methods of making a cellar. Mr. T. F. Bingham, of

Mr. Bingham's Cellar that Resembles a Cistern.

Farwell, Michigan, has a cellar that has been compared to a *cistern*. The walls are made sloping, and then plastered over very heavily with cement, after the manner in which cisterns are sometimes made. Over the cellar is laid a floor covered several inches with dry sawdust, while a roof keeps all dry. Mr. Bingham is a believer in having fresh air for the bees, even though they use only a small amount, and he has a ventilator 17 inches square running up through the ceiling and roof. Mr. Bingham also finds this ventilation of great help in keeping the bees quiet during the first warm days of spring, before he considers it late enough for their removal.

Some parts of the country are too low and level to allow the building of a cellar below the surface of the earth, when some sort of a structure above ground is the only resort. Some of these above-ground cellars have double walls built of brick, others have walls of stone, and still others are made of cedar or pine logs after the manner of a log house, and the whole structure then covered with earth. A cellar thus surrounded by earth is almost as thoroughly proof against the changes of temperature, as though built under ground.

Having briefly considered cellars, let us come back to the subject of temperature; and, by the way, I am certain that I can do no better than to quote a few paragraphs upon this subject from an article contributed by Mr. R. L. Taylor to one of the early numbers of the Bee-Keepers' Review. Among other things, Mr. Taylor said: "I think it a truth not to be forgotten that no one can determine, except approximately, the best temperature for bees in another's repository. The condition of the bees as to numbers, the warmth and ventilation of the hive, the character of the hives, and the state of the repository as to moisture, have each to be considered in deciding upon temperature.

Of course, the bee-keeper cares nothing about the temperature in *itself*; what he is interested in is in knowing what the condition is in which the bees pass the winter with the least loss of vitality. In what manner temperature affects this condition is really a subsidiary question. If we could agree upon the primary question, I think there would be little difficulty in solving the subsidiary one.

What are the distinguishing marks of the condition most desirable for the well being of the bees?

We know that at the beginning of their season of rest, bees cluster closely, and we know that so strong is this instinct that this state, late in the fall, continues in a temperature that at another season of the year would cause extreme activity. There is no doubt that this is the state best suited to the preservation of the physical

Cellar Built of Logs and Covered with Earth.

powers of the bees. Labor, activity, anxiety, are wearing to mortal flesh. To live long, one must live slowly. We wish our bees to have the same degree of physical vigor in April which they possess in November. I would emphasize the adverb in the phrase 'cluster closely,' in using it as an earmark of the condition desired. The quietness sought should be a quietness to the eye, and not to the ear alone. The right cluster is knit together, and the individual bees thereof only aroused to full consciousness by positive disturbance. Bees in a loose cluster, or spread through the hive, often make little sound when wearing themselves out by premature brood rearing or by over feeding. How does temperature affect the desired condition?

Most bee-keepers know that temperature below a certain point causes activity among the bees on account of the necessity they feel of keeping up the warmth of the cluster by exercise, in order to prevent themselves sinking into such a degree of chilliness that they shall no longer have the power to resuscitate themselves; and all know that as the period of rest lengthens, the bees become more and more susceptible to a high temperature, and are very likely to be pushed by it into unseasonable activity. Again, the temperature may be so low and so long continued that, notwithstanding their efforts, they perish either of cold or starvation.

Of course, the temperature that determines the welfare of a colony is that within its own hive, so it becomes very important in fixing the temperature to consider the strength of the colonies, and size, warmth and ventilation of the hives. A temperature that would enable a weak colony to winter safely would almost surely greatly injure a strong colony in a hive of like size and condition, unless its stores were of good quality, and *vice versa*. Weak colonies should be protected by contraction and a closer hive—the stronger given more ventilation. A moist atmosphere conveys away animal heat much more rapidly than a dry one, so that the best temperature in one cellar might vary many degrees from that which would be best in another.

I have no doubt in my mind that, with stores which are exceptional, every normal colony would winter well in any ordinary bee-cellar, where the temperature ranges from 32 to 50 degrees, Fahrenheit, and that we err when we attempt to make successful wintering turn upon anything aside from food; still, no doubt the temperature may be made to assist the bees in contending with the distresses arising from the unfit food. Warmth makes the discomfort of their diarrhoetic disease less unbearable. In a low temperature, bees afflicted with diarrhoea soon perish miserably. So, for bees thus diseased, I would provide a high temperature; say about 50 degrees, thereby enabling the dying to leave the hive, the diseased to void their excreta outside the cluster, and the well to make a more courageous fight for life.

I need scarcely add anything more upon this part of the subject, and shall only say farther that, in my own cellars, where the air is neither very moist nor very dry, and where there are no draughts, I consider a temperature of 40 to 44 degrees the best for good colonies in hives from which the bottom boards are entirely removed. If the bottom boards are not removed, I think that five degrees lower would be about equivalent.

In order to have the temperature as desired, it becomes important to have one's bees in a repository of which the temperature is nearly independent of outside changes. This is, I think, secured far more satisfactorily by having the repository entirely, or, at least, very largely, below the surface of the earth."

As the temperature is higher at the upper part of a cellar, the weak colonies should be placed in the topmost tier of hives.

It has been urged that, as spring approaches, and breeding begins, the temperature of the cellar should be raised. With a large number of colonies the increased activity would, of itself, have a tendency in this direction. If there are only a few colonies, artificial means of raising the temperature are sometimes employed. Some have used oil stoves in the hatchway of the cellar; others have warmed the air with wood or coal stoves. If an oil stove is used, there ought to be a metal hood over it, and a pipe connecting with a stove pipe in the room above, or else with the open air. Of course an oil stove can be used without such an arrangement, but it overloads the air with the gases of combustion. I mention these make-shifts with something akin to reluctance, as I feel that the proper way to do is to have a cellar so constructed that there will be no necessity for their use.

Mr. H. R. Boardman, who has had much successful experience in wintering bees in cellars, prefers to have a bee cellar with two apartments, in one of which is a stove. If he ever finds it necessary to resort to artificial heat, he warms the air in the ante room, and then admits it to the room. In the use of artificial heat he does not find it necessary to employ it constantly, or every day; in fact, he says that the best results are secured by giving the bees the benefit of a summer temperature for a short time once a week, and then letting them alone. They will, after being warmed up, become quiet in a short time, and remain so for several days, and no serious results may be apprehended from cold, if in a frost-proof cellar.

Wintering Bees in a "Clamp."

There is still another method of securing the proper temperature for wintering bees, aside from that of packing them in chaff, or putting them in the cellar, and that is that of burying them in "clamps," as they are called, the same as potatoes and apples are buried in pits. A long trench is first dug a little wider and deeper than a hive. In the bottom is placed a layer of straw, then two pieces of scantling upon which to set the hives. Rails, fence posts, or any kind of supports, are then laid over the hives, and covered with straw upon which the earth is thrown to a sufficient depth to exclude the frost. Sometimes ventilation is given these clamps, but it does not seem to make any material difference whether they are ventilated or not. It does make a difference, however, in regard to the soil and situation. In a sandy or gravelly knoll, where the water will never stand, the successful wintering of the bees is almost assured. In heavy clay, the loss of the bees is equally certain. I say this from numerous experiments. Bees in a clamp, in the right kind of soil, in a good condition, winter equally as well as in a cellar, sometimes it seems as though they winter better, and the only possible objections to this method are the labor and untidiness.

Care of Bees in Winter.

IF THEY were properly prepared for winter the preceding autumn, given plenty of good stores, properly protected out of doors, or placed in a cellar or other repository having the proper temperature, and precautions taken against depredation by mice, bees require almost no care in winter.

No bee-keeper worthy the name will allow his bees to go into winter quarters short of stores. They ought, at least, to have enough to last them until the first warm days of spring, when they may be handled upon their summer stands, and fed if necessary.

However, if by any hook or crook, bees *have* gone into winter quarters short of stores, and there are fears that some of them may be starving, it is better that they be examined and fed if needed, even though the task may be unpleasant. There need be no hesitancy in thus disturbing the bees for fear that it may do them some injury, for, as a rule, it will not.

Probably the best method of feeding a colony of bees in winter, is to give them a frame of honey. If no honey is available, and some of the colonies *must* be fed, the best substitute is candy made from granulated sugar. Put in sufficient water to dissolve the sugar, then boil the syrup until it will harden in cooling. To learn when to remove the candy from the stove, take out a spoonful every few minutes, and allow it to cool. As soon as it begins to show signs of hardening, draw the vessel containing it to the back of the stove, where the heat is less. Watch it carefully and try it frequently. As soon as it is sufficiently hard, remove it from the stove, and pour it into shallow dishes to cool. Be careful not to get it too hard. If it is hard enough to retain its form when placed over a colony of bees, that is sufficient. A thin cake of such candy laid directly upon the frames over a colony of bees, and then the whole top of the hive covered with a piece of enameled cloth, and two or three thicknesses of old carpet over that, will enable the bees to "hold the fort" as long as the candy lasts. If, for any reason, it is impossible, or undesirable, to place the candy in this manner upon the tops of the frames, the candy may be "run" directly into empty brood frames, and the frames hung in the hives adjoining the bees.   To fill a frame with candy, lay it upon a smooth board with a piece of paper under the frame, and pour in the candy, after first waiting for it to cool until it is as cool as it can be, and yet be made to "run."   To keep the frame down close to the paper, so that the soft candy will not run out while cooling, tack the frame down with some nails just long enough to hold the frame down nicely, but not long enough to make it difficult of removal.   If a frame full of candy is more than a colony needs, a less amount may be given by tacking a crossbar in the frame, part way up from the bottom, and filling the upper space only with candy.

Mice sometimes do some little damage, both to colonies wintered indoors, and those in the open air.   This damage is confined principally to that of gnawing the combs.   If bee-keepers would only remember that bees can pass through a space that is less than ¼ of an inch, and that a mouse needs a space nearly twice this, it would seem that there need be no trouble in keeping mice out of doors. Simply contract the entrance until it is only ¼ of an inch the narrowest way, and no mice can enter.   This should be done quite early in the fall, as cool, frosty nights often drive the mice into the warm retreat to be found inside a bee hive.   When bees are wintered in the cellar, many bee-keepers practice raising the hive about two inches from the bottom board; others remove the bottom board entirely.   This allows plenty of ventilation with scarcely any escape of heat.   All dead bees and rubbish drop down away from the cluster of bees, where they dry up instead of becoming mouldy and rotten from contact with the warmth and moisture of the cluster.   If a colony *does* die, the combs are left dry and clean, instead of being stuck together with a mass of damp, moldy, rotting bees.   All who have tried raising hives in this manner are enthusiastic in its praise; but it will be seen that this plan gives the mice, if there are any in the cellar, free access to the hives.   The remedy is to trap the mice, or poison them.   For the latter purpose I have found nothing better than equal parts of flour, white sugar and arsenic, mixed, and placed in shallow dishes in different parts of the cellar.

Unless the cellar is well under ground, where it is well beyond the influence of the outside temperature, it is well to keep watch and not allow the temperature to run too low in *protracted* cold spells. A lamp stove, burned all night in a cellar, will raise the temperature several degrees.   During the fore part of winter, a low temperature is not so dangerous as it is towards spring, when brood rearing may have commenced.   From 35 to 45 degrees will answer very well until towards spring, when it ought not to be allowed to go below 30 degrees, and may with safety go as high as 48 or 50 degrees. So long as the bees remain quiet, I should not disturb them with artificial heat. If the cellar becomes *too warm* in the spring, before it is time to remove the bees, it may be cooled down by carrying in ice or snow, or the windows and doors may be opened at night and closed in the morning.

Years ago, many bee-keepers practiced taking their bees from the cellar, if there came a warm day in the winter, and allowing them to fly, returning them again to the cellar, but this practice has been pretty nearly abandoned. If the bees are in a quiet normal condition, it often rouses them, and sets them to breeding in mid-winter, which is far from desirable. Rapid breeding late in winter, or very early in the spring, is decidedly objectionable; nothing so quickly wears out bees as the rearing of brood; and the more unfavorable the conditions, the greater the wear. It is better that the bees should remain quiet until warm weather furnishes the most favorable conditions for brood rearing, when the same expenditure of vitality will produce two bees instead of one. Therefore, don't allow a warm day or two in the winter to tempt you to the removal of the bees from the cellar. Wait until the snow is gone, and there is occasionally a day warm enough for bees to fly, then take them out to remain permanently. On the other hand, nothing is gained, and much may be lost, by leaving the bees in the cellar until *late* in the spring. Many claim superior advantages for out-door wintering, asserting that the colonies build up earlier in the season. They won't if the bees are taken from the cellar early enough; and, certainly, it requires no argument to show that bees successfully wintered in the cellar are better able to bear the rough weather of spring than bees that have endured all of the rigors of the entire winter out of doors. In most of our Northern States the main honey harvest comes early in the season, and to secure this harvest there must be a goodly number of field workers *at the right time*, and the eggs from which these workers are produced must be laid several weeks previous to the opening of the harvest, hence the element of *time* is an important factor, and nothing stimulates a colony in a healthy manner, and sets it to brood rearing, as does a flight in the open air, even if nothing is brought in. Hence it will be seen that early removal from the cellar gives the bees largely the advantages of both out-door and in-door wintering.

There is no danger of the cold injuring the bees when they are *first* removed from the cellar, the trouble comes from late freezes coming after two or three weeks of fine weather. At this time the

Colony of Bees Protected With Building Paper.

combs are filled with brood, the cold drives the bees into a compact cluster in the center of the hive, and all of the brood outside of this perishes. All of this loss may be avoided by giving the bees some sort of protection after taking them from the cellar. First see that each colony has a queen and plenty of stores, and then protect it. This spring protection need not be an elaborate affair. A sheet of tarred building paper folded down over the hive, and fastened at the lower edges by tacking on strips of lath, will answer every purpose, while it costs only three cents, and can be put in place in less than five minutes. This makes a covering that is both wind and water-proof, and will absorb every particle of the sun's heat, but, more important than all this, it will save the loss of brood and weak colonies if there comes a "squaw winter" in the month of May.

If spring protection is so important that it is advisable to give it after taking the bees from the cellar, it may be asked, why not practice out-door wintering, then winter-protection will answer for spring, and the expense of a cellar, and of carrying the bees in and out, will be avoided? In the first place, the saving of stores in cellar-wintering will pay for the expense twice over; and, in the next place, and of far more importance, it is only by the cellar method that the wintering of bees, in a cold climate, can ever be reduced to a perfect system. By a selection of natural stores, or, better still, by using sugar, we can secure uniformity of food, but it is only in the cellar, or special repository, that uniformity of temperature, at a desirable point, can be maintained.

Carrying the bees from the cellar is not a very agreeable task, aud most of bee-keepers make it much worse by attempting it upon such a warm day as to set the bees fairly crazy the moment the out-door air strikes them. It comes into the cellar and sets the bees to flying, and often there is a general mix-up in the yard by the bees of one colony joining with those of another in full flight, and following them into their hive. To avoid these troubles, some bee-keepers carry their bees out in the night, when the indications are that the following day will be fair. If the bees have *wintered perfectly* and are *quiet*, all of these annoyances and losses may be avoided by carry-ing out the bees upon a day so *cool* that the *bees will not think of flying*. This idea that bees must fly the moment that they are taken from the cellar is one of those old notions that is a *notion*, and that is all. If bees have to wait even a week or two after being placed upon their summer stands, before having a flight, no harm will come as the result, *providing* they have not wintered poorly, and are so anxious for a flight as to leave their hives when the weather is so cool that they will never return.

Conclusion.

IN CONCLUSION let me say that the writing of this book has been largely a labor of love, of hope, of a desire to benefit bee-keepers; to arouse, encourage and inspire them, and lead them to adopt better methods.

I wish them to have broader views of their occupation, to look upon it as a *business*, to stop "fussing" with a *few* bees, to get rid of other hampering pursuits, to branch out and keep enough bees to employ all of their time, energy and capital.

No more ennobling pursuit exists than that of bee-keeping. It is the poetry of agriculture. It is uplifting and inspiring, health-giving and useful, fascinating and profitable. It improves the mind, trains the eye and hand, cheers the heart, and fills the pocket book. Knowing all this as I do, it pains me beyond expression to see it maligned and belittled, to hear it called a precarious and uncertain pursuit, one that must be tacked on to the tail of something else, when I *know* that, rightly managed as a specialty, there is no rural pursuit more safe, pleasant and reliable.

If this book does no more than to show the possibilities of advanced bee culture, if it only succeeds in giving the doubting Thomas courage and faith to rid himself of all encumbrances, and then press on to success with bees alone, if it helps to change bee-keeping from a hap hazard, happy-go-lucky side-issue to the dignity of a reliable business, it will not have lived in vain.

Reader, the time has come to say good bye. To me it seems like parting from a dear friend; and, in closing, let me say that I would be delighted, at any time, to receive criticisms, suggestions or queries regarding any of the topics mentioned in this book.

Index.

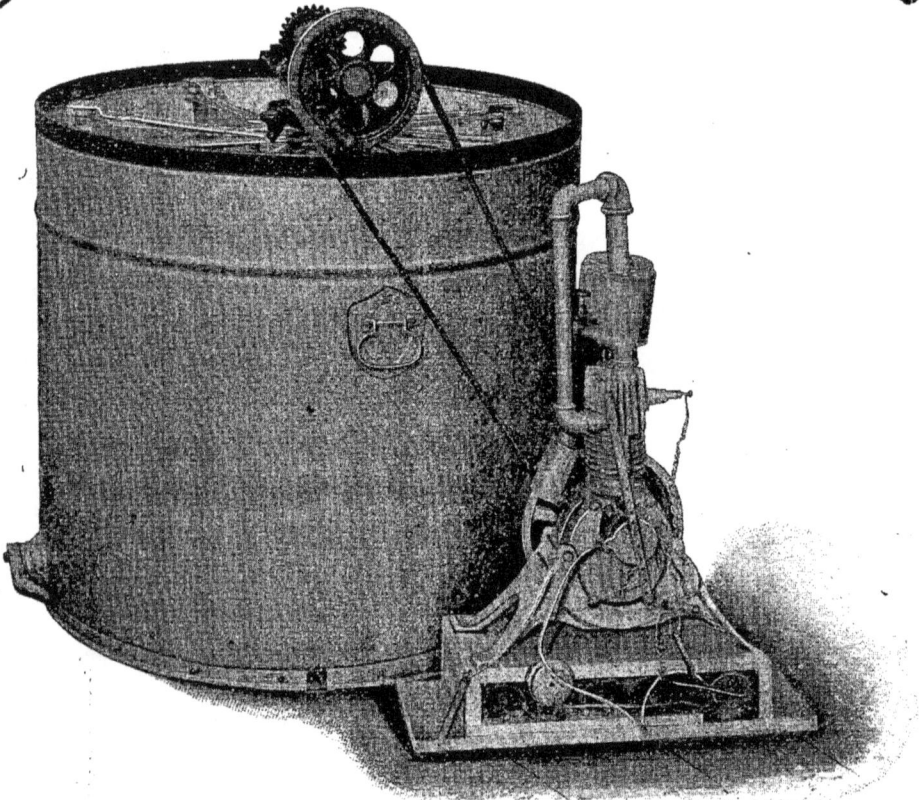

The Bee=Keepers' Review

Is most emphatically the specialist's journal. For 15 years its editor made his living in the apiary, and he knows the needs of this class of bee-keepers. Instead of using space for "hints to beginners," which are all right in their place, the Review turns its attention to the unsolved problems of advanced bee culture. Some of the best best bee-keepers in the country, those who have managed large numbers of colonies, and made money in so doing, describe their methods in the Review. These men have succeeded. They have made money. They can point the way for others. The man who has kept bees several years, who knows the A B C of the business, is now interested in learning systems, methods and short-cuts, that will enable him to spread out and "keep more bees," and make some money; and no journal is doing more for this class of bee-keepers than is being done by the Review.

The Review is an illustrated, 36-page monthly at $1.00 a year. If you wish to see copies before subscribing, send ten cents for three late, but different issues, and the ten cents may apply an any subsription sent in during the year. A coupon will be sent entitling the holder to the Review one year for only 90 cents.

W. Z. HUTCHINSON, Editor and Proprietor **FLINT, MICH.**

BEE SUPPLIES

Best equipped factory in Wisconsin. A large stock and variety of everything needed in the Apiary. Best goods at the lowest prices, and prompt shipments. I want every bee-keeper to have my free illustrated catalogue and read the descriptions of the Celebrated Cedar shingle roof bee hive cover.

Write at once for catalogue.

A. H. RUSCH,

Manitowoc Co.   Reedsville, Wis.

Make Your Hives

BEE keeping is busy work in the summer-time; but the winter brings a leisure that many more bee-keepers might profitably employ in making needed hives, supers or shipping cases for another year. Power and expensive machinery are not needed; simply a cosy little shop and a foot-power saw are all that are needed. When a bee-keeper realizes all this, there is no question as to what saw he shall buy; it is made at the factory of **W. F. & JNO. BARNES CO., Rockford, Ill.** The author of this book has used one of these machines, and has no hesitation in saying that it is all that is claimed for it. Any one who buys a machine, and is not entirely satisfied with it, has the privilege of returning it and having his money returned. One thing more; there are attachments, such as a scroll-saw, a boring attachment, etc., that can be added at a small cost. Send for catalogue.

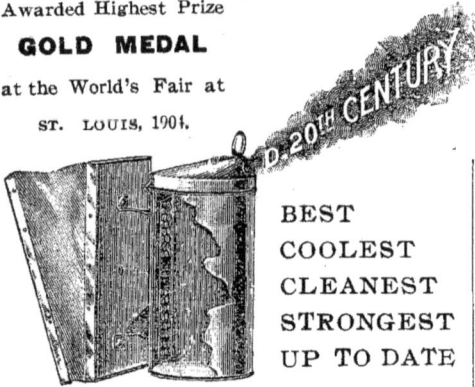

Honey Cans and Pails

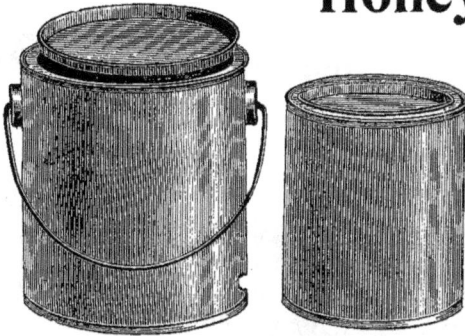

FRICTION TOP STYLE

Made by the

AMERICAN CAN CO.

New York
Chicago
San Francisco

These cans furnish ideal Honey Containers, in that they provide a large opening for filling or emptying, and are readily opened and closed. Having no projecting caps, they pack conveniently for shipment.

Capacity	Diameter	Height
1 lb. Cans	2 11-16	4
2 " "	3½	4¾
2½ " " short qt.	4	4 9-16
3 " " full "	4 3-16	4⅞
5 " Pails short ½ gal.	5 1-32	5 11-16
10 " " " 1 "	6 3-16	7½
6 " " full ½ gal.	5 3-8	5 7-16
12 " " " 1 "	6 17-32	7 3-16
24 " " " 2 "	8¼	9⅛

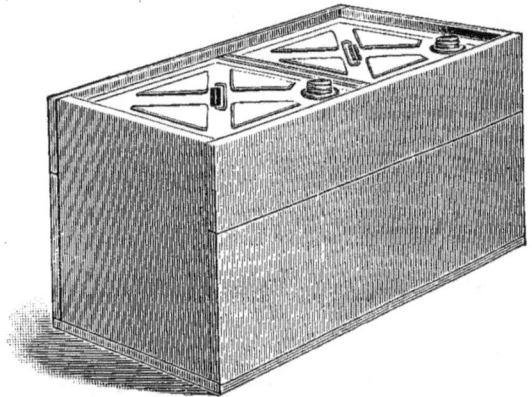

These cans and pails are made at Maywood, Ill., near Chicago. All sizes larger than three pounds have wire bail handles. Special sizes can be furnished if ordered in large quantities.

60 Pound or 5 Gallon Square Cans

1¾ inch Screw Nozzle

DOUBLE CASE

Cans are provided with wire handles as shown in cut.

Cases have ⅞-inch ends and ⅝-inch paritions, balance ⅜-inch lumber. Ends of cases have cut handles.

We also furnish same style in single cases.

Shipping weight of single case 8½ pounds.

Shipping weight of double case, 15 pounds.

60 pound or 5 gallon round cans, wood jacketed, wire bail handles, 3-inch screw opening, independent jacket. Inside tin can easily removable from jacket to admit of contents being heated to facilitate pouring. Prices on all the above quoted upon application stating quantity wanted.

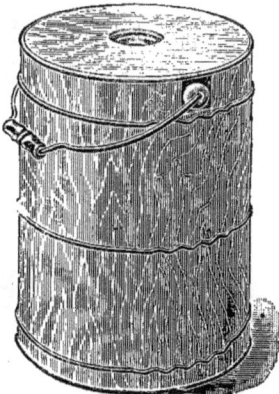

AMERICAN CAN COMPANY

New York   Chicago   San Francisco